Shailesh (Sky) Kadakia

True Physics of Light,

Beyond Relativity

Revealing the Magic and Mysteries

Behind the Creation of the Universe

With 40 illustrations of which 29 in Color

MWP PRESS

MATRIX

True Physics of Light, Beyond Relativity

Shailesh R. Kadakia, MSEE, 602 Suburban Court, Apt. 3, Rochester, NY 14620 USA

Publisher:	MATRIX WRITERS & PUBLISHERS
Distributor:	AtlasBooks Distribution
Registration Services:	Kelly Sabetta
Website Creator:	Patti Jacobs
Editorial Production:	Maureen A. Nielsen
Illustration Designer:	Shailesh Kadakia
Copy Editor:	Shailesh Kadakia
Permissions Editor:	Shailesh Kadakia
Cover Designer:	Shailesh Kadakia
Cover Image Front:	Copyright 2010 MATRIX WRITERS & PUBLISHERS
Compositor:	MATRIX WRITERS & PUBLISHERS

Library of Congress Cataloging-in-Publication Data

Shailesh R. Kadakia, MSEE
True Physics of Light, Beyond Relativity/Revealing the Magic and Mysteries
Behind the Creation of the Universe/Shailesh R. Kadakia
Include bibliographical references and index.

Library of Congress Control Number: 2010900129

ISBN-13: 978-0-692-00707-5 (Soft cover, Alkaline Paper)
ISBN-10: 0-692-00707-5

Printed on acid free paper

MWP Press is an imprint of MATRIX WRITERS & PUBLISHERS

Printed in the United States of America
9 8 7 6 5 4 3 2 1

ISBN 978-0-692-00707-5 MATRIX WRITERS & PUBLISHERS, Rochester, NY

True Physics of Light,

Beyond Relativity

Dedication

To my parents,

Lilavanti R. Kadakia & Rasiklal S. Kadakia

Thank you for your constant encouragement and support.

– Shailesh Kadakia

Preface

An energy source, such as light, can be seen by humans yet cannot be otherwise sensed. It is known to have no material mass yet has revealed visible and physical effects of mass through several physical phenomena, such as photoelectron emissions and others. This strange nature and behavior of light energy waves has made it the most poorly understood energy source of nature. Often in the past, many scientists have erroneously modeled the dual behavior of light, sometimes as a particle and sometimes as a wave. After reading this book, you will be convinced that light energy waves strictly behave as waves, and that all of the phenomena and results of experiments over the past two centuries, to prove that light occasionally behaves as particles, were incorrect. Another source of energy is the force of gravity whose cause is not known to date. In this book, we have attempted to analyze the root causes of the force of gravity, among different celestial objects in the universe. Another puzzling question is, at what speed the force of gravity propagates? Albert Einstein had suggested that in vacuum, the propagation speed of gravity is the same as the speed of light waves. According to the postulates of the special theory of relativity, the speed of light in a vacuum is constant **c**. The speed of light varies according to the frame of reference, which is in-concert with the varying speed of light (VSL) theory proposed by the Portuguese physicist, João Magueijo, a research fellow in theoretical physics at Cambridge University. Therefore, it is imperative to answer the question, at what speed the force of gravity among different objects propagates. The speed of light is no longer constant in accordance with the **Skylativity®** theory.

The main intention of this book is to correct the basic understanding of the fundamental ideas regarding the nature of light, the special theory of relativity, and the general theory of relativity. The book is primarily written for first-year college students who are considering physics major. It assumes competence in high school algebra, geometry, and calculus. Since our emphasis is on improving the conceptual understanding of physics, and not to understand mathematical rigors, we have discussed analytical expressions with simplicity. Another purpose of this book is to determine an ideal theory of everything that will tie the creation of the universe, which contains huge celestial objects, such as galaxies, super nova, and nebulae, to the tiniest fundamental particles, such as quarks and fermions. Also, we shall utilize the concepts of the **Skylativity®** theory to marry classical mechanics with quantum mechanics and create a universal unified field theory (UUFT) of everything.

The content of this book is organized as follows:

In Chapter 1, we begin with a discussion on the nature of light and we are very critical about its behavior as a wave and not as a particle in some events, as quoted by Einstein and others. Then, we touch on some of the events that our predecessors thought were occurring as a result of the bending of light by the force of gravitation. A detailed examination of the analysis of events will convey that it is impractical to predict that light waves are bending by the force of gravity. Next, we explain why the conclusion that light behaves as a particle, based on experiments performed in early 1900's, through modern times, may be incorrect. We demonstrate this by analyzing the sets of experiments of the photoelectric effect, short wavelength limit X-rays, the Compton Effect, and Michelson's interferometer. We substantiate the wave theory of light by explaining Doppler's shift effect and the rotation of a fan when its thin blades are coated with semiconductor paste and are exposed to light. Also, we point out that the concepts of time dilation, introduced by Einstein and Lorentz, are incorrect as a result of the wrong conclusions drawn from the analysis and results of Michelson's apparatus. In this chapter, we successfully indicate that some of the ideals proposed by Einstein, and subsequently promoted by Lorentz and others, may require further analysis.

In Chapter 2, we describe the basic differences between the physical properties of a wave like entities, and particles. Then, we explore the speed of the propagation of light and electromagnetic waves. We emphasize that the speed of light and electromagnetic waves is variable and it should vary differently in medium with different refraction coefficients. Also, it should alter if the frames of reference are moving at different speeds. Then, we introduce the concept of the true speed of light and formally define it, based on the distance between the speed measuring instrument and the source of light as invariant over the arbitrarily small time interval. We explain the importance that not all of the energy waves of the electromagnetic spectrum should be travelling at the same speed. Therefore, it may be imperative that the electromagnetic spectrum table should be split into three or more tables. We briefly discuss the strengths of the interactions of weak, strong, and residual forces on interacting particles and compare them over the range of the distances they encounter. Next, we analyze the expressions for the energy released during a nuclear radiation and explosion event and in general, the method for specifying the energy content of the light waves. In our analysis, we prove that the energy released during a nuclear explosion event, computed by Einstein's equation $\mathbf{E} = \mathbf{M} \times \mathbf{c}^2$, estimates the energy liberated in the excess amount rather than the actual value. Also, we explain that the major source of the energy released, is from the liberation of the binding energy of the nucleons at the core of the atoms of the radioactive matter and not the consumed mass of the matter, as explained by Einstein, in a nuclear explosion device, when detonated. We provide the correct analytical expression for the released binding energy. In this chapter, we also choose to discuss the topic of the temperature profile of the prime energy source, the sun in our solar system. We want to direct our attention to the fact that the temperature calculations predicted by the current techniques have provided large values for the surface and core temperature of the sun. These numbers are unrealistic and other methods are needed to confirm the results of the measurements to date. In particular, this is true for the 15M°K estimate for the core temperature of the sun. To predict the temperature profile for the sun's core, we have proposed a different solution.

In Chapter 3, we start explaining the theory of relativity, as postulated by Einstein, which has been widely accepted for a little over the past century. Here, we want you to be familiar with and refresh your memory about the basic concepts of the general and special theories of relativity, first introduced by Einstein. We shall refer to his theory as the E-theory, from the name of its inventor, Einstein. Also, we shall use a similar name to abbreviate and refer to the postulates of special and general **Skylativity®** theories introduced by us, as K-theory. In the next section, we state postulates of the special and general **Skylativity®** theories. We highlight the differences between the postulates of E-theory and K-theory and explain the implications of the new theory to the applications of radio astronomy and cosmological measurements. Next, we point out, that on the basis of the new theory, Lorentz expressions for length, mass, and time computations for different inertial systems, are not required. Then, we discuss the modifications required in Maxwell's field equations, to reflect the variation of the speed of light among different frames of reference. Finally, we describe the changes required in the solution of Einstein's field equation that take into account the varying speed of light **c**.

In Chapter 4, we begin with a discussion of the limitations of Einstein's general and special theories of relativity, when applied to the measurement of length, time, and mass. We describe an example of a common incandescent light bulb to show that light energy can be generated when a finite amount of electrical energy is supplied. By showing this, we are stating that with a finite source of energy, light photons from a tungsten wire filament can be accelerated from zero speed to light speed. This contradicts the claims of the earlier theory that an infinite energy supply is needed to attain the speed of light. In Section 4.2, we indicate that, in the same example of the light bulb, when a filament emits light, it does not lose any mass. Therefore, for every event where the light energy is released, the popular energy to mass conversion relation, $\mathbf{E} = \mathbf{M} \times \mathbf{c}^2$, does not hold well. In Section 4.3 we explain that if we believe for an object, a different measured mass value for a different inertial system, we have to infer that the speed of light should be different in those frames of reference. This is true because the mass in both reference systems possesses the same rest mass energy. To resolve this conflict of mass variations for masses of moving objects, we introduce a concept that total mass consists of a real rest mass component and an energy mass component. The distribution of the rest mass component supposedly affects the center of gravity. The orthogonal imaginary component, the energy of mass, may affect the future position of the center of gravity after time $\boldsymbol{\delta t}$ is known as the dynamic mass. In Section 4.6, we propose that time dilation is a fictitious concept that was introduced by Einstein and others. His skewing of the time dimension does not make sense. Therefore, the time dimension is invariant and measured time will always be the same if clocks in the two inertial systems are truly identical. In the final section 4.7, we show that the stellar parallax distance measuring method would introduce a vast number of errors, if we assume that the light rays arriving from other stars, when passed by the sun, will be deflected by the force of gravity from the sun. Hence, it makes sense to state that light rays do not bend by the force of gravity.

We firmly believe that the Lorentz transformation equations for mass, length, and time measurement, for different inertial systems, are not needed. Therefore, in Chapter 5, to investigate a hypothetical situation, we explore the effect of the variable speed of light on the mass, length, and time measurement, using the current Lorentz transformation. We analyze the values of mass, length, and time measurements for two different scenarios, approaching systems and receding systems. In the infinite universe, those two scenarios play a more important role in the space coordinate transformation than the linear movement in the X, Y and Z directions for two or more systems.

In Chapter 6, we highlight the benefits derived when the new postulates of the **Skylativity®** theory are applied to modern day astronomy and space science. In the first section, we describe the ways that create surplus funds and resources. We convince you that it may not be essential to invest funds in the construction of huge super colliders. At present, a significant amount of resources are spent to build proton accelerators to determine if the speed of light is achieved by a particle. The postulates of the **Skylativity®** theory state that the speed of light is achievable by particles, therefore, the financial and manpower resources may be saved by not applying them toward the construction of huge super colliders in the future. In Section 6.2, we stipulate that the measurement scales for mass, length, and time units are universal and constant among different frames of reference. This approach avoids the complex formulation of the Lorentz contraction of length and time dilation, while computing the coordinates in different inertial systems. In Section 6.3, we address the decay rate of very weak interacting neutrino particles from the sun. Scientists believe that the lifetime of these particles is very short, so, they will disintegrate before they reach the Earth's atmosphere. The particle survives because the decay rate is slowed by the time dilation factor computed according to the Lorentz formulation. We believe that time dilation is a virtual effect and should not affect the decay rate and disintegration of the neutrinos. In Section 6.4, we propose that the discovery of quarks allows us to develop future weapon systems with enormous power, similar to the fission of ^{238}**U** into ^{235}**U**. Alternatively, the controlled triggering mechanism that smashes protons and neutrons into quarks may be applied to design power plants. These power plants have the potential to generate a vast amount of energy source from the release of the binding energy of quarks. We shall call this proton and neutron power plants. In Sections 6.5, 6.6 and 6.7, we discuss the future of the space program and suggest ideas for the design of spacecrafts that travel at fractions of the speed of light.

In Chapter 7, we develop the universal unified field theory (UUFT) that integrates the effect of gravity from macroscopic objects, such as celestial stars, galaxies, and nebulae, with the strong forces of particles in the standard model which deals with microscopic particles. This has been a huge challenge in the past because gravity is found to be a very weak interacting force, as compared to the charge and spin momentum forces with strong interaction effects within nucleus of tiny atoms of particles. Next, we analyze the reasons for gravity. Every object in the universe projects a force of gravity on another object because both objects possess momentum and potential energy associated with each other. As per our explanation, the force of gravity exists among any two objects that have real mass and a static location for the center of gravity. Therefore, the force of gravity from large celestial objects does not have any effect on the trajectory of wave entities, such as light rays and electromagnetic radiation energy waves. In section 7.3, we briefly discuss time travel, which is a fictitious concept. Time travel only exists in your imagination because time dimension cannot be retraced. For instance, the conversion of hydrogen into helium atoms, through the thermonuclear burning process on the sun, is an irreversible process. This implies that, at the end of the life of the sun, the death of our civilization is imminent. Also, no power on earth could ever change the rate or speed at which the sun orbits on ecliptic to the center of galaxy. In Section 7.4, we discuss the reasons why the weather forecasts are not accurate at all times.

In Chapter 8, we focus on black holes and the origin of the universe. In Section 8.1, we explain that black holes do not have super gravity. Many scientists have claimed that black holes are massive with a super gravitational field in which light is trapped. We believe that light does not escape from black holes because it is absorbed. We provide a formal proof of our theory. Also, we believe that the universe is neither expanding nor contracting because the space of the universe is boundless. If we state that the universe is expanding, it implies that we know there is something outside the limit because it must expand into space that was either occupied previously or created by the expansion. There is no evidence which proves that such an expansion is observed inside of our galaxy. It is not obvious how the selective expansion of the universe could occur outside our galaxy. Our prognosis about the observed red-shift of celestial objects, such as other galaxies and supernova is correct because they are moving away (receding) to maintain the balance between gravitational effects and centrifugal force. Thus, it is essential to redefine Hubble's constant in his law for the three dimensional movement of celestial objects. Next, we suggest that the mapping of the sky should be partitioned into past, present, and future universes, according to the separation of celestial objects from our earth and the solar system. In Section 8.5, we discuss the ultimate fate of our solar system after all of the hydrogen is transformed into helium at the sun's core through the thermonuclear burning process. In Section 8.6, we explain why the planets Venus, Uranus, and Pluto, rotate from the east to the west on their axis instead of the west to the east motion of the Earth, Jupiter, Saturn, and other planets. In Section 8.7, we explain why the orbits of comets are asymmetric. In the final section of this chapter, we look at some of the advances in modern physics, such as the String theory and new dimensions.

In Chapter 9, we discuss the ways to bridge the gap between the classical Newtonian mechanics and quantum mechanics. In Section 9.1, we revisit the outcome of Young's double-slit experiment and the behavior of a quantum particle electron in a shell orbit, by applying the principles developed by Erwin Schrödinger. In Section 9.2, we analyze the discrete model for radiation from the surface of a black body invented by Max Plank. We show that the quantum of energy possessed by the light wave, Plank's constant **h**, is inherited from the parent particle. Therefore, **h** is property that is associated with the quantum particle electron and not the fictitious particle photon. In this chapter, we establish that the quantized model to characterize black body radiation, from Plank, should not prohibit us from proving the wave as the only model for light and radiation energy. In Section 9.3, we discuss an important contribution to quantum mechanics, from a somewhat less recognized physicist, Paul Dirac. His ideas were instrumental in the prediction of the existence of the complementary particle, pair proton, anti-proton, electrons, and positrons. Also, his equations validated many different concepts and theories, such as Pauli's theory and the hole theory of atoms. In Section 9.4, we address the main objective of this chapter, to connect classical mechanics and quantum mechanics. We achieve this objective by explaining the operation of quantum devices, based on Schrödinger's equations. In Section 9.5, we discuss the practical application of quantum mechanics by looking at an example of the scanning tunneling microscope. In Section 9.6 we describe the applications of quantum mechanics as a solution of complex problems, such as finding material that exhibits super conductivity at and near room temperature.

Further information and details about the topics discussed in this book can be obtained at the web site **http://www.Matrixwriters.com**.

Acknowledgment

Mr. Kadakia would like to thank his daughter, Crystal, for her assistance, advice, and technical discussions. We especially acknowledge the efforts of Dr. Jaffer Amin, Engineering Fellow from Raytheon Corporation for proof reading the book for technical accuracy. We admire him and Sam Tang for spending time on late nights discussing the validity of concepts presented in this book. Further, we wish to extend our thanks to Maureen Nielsen reviewing and editing the book, making it more readable. We appreciate the book registration services of Kelly Sabetta for making this project successful. We would like to thank Patti Jacobs for designing and maintaining our Web site. We are extending our gratitude to Erin Bushart, Manager at a local FedEx Office retail store for reviewing the book and checking English language usage. Further, we thank, Mr. Kadakia's cousin, Kirit Dharia, for providing the photograph of Albert Einstein and Rabindranath Tagore from newspaper clips that he saved and sent to us. Also, we extend thanks to Mr. Kadakia's nephew, Meet Kadakia, for useful remarks.

Our sincere thanks go to President, Dr. J. J. Rawal and Scientist Viren Dave, both of the Indian Planetary Society, for meeting with Mr. Kadakia during his trip to India in the fall of 2009. We admire their valued comments and discussion on the subject of relativity and current technology. Also, we appreciate the valuable comments from Emeritus Professor, Dr. Jayant V. Narlikar of Inter University Centre for Astronomy and Astrophysics (IUCAA), Pune, India. Further, we appreciate the suggestions from Dr. Steven Manly of the University of Rochester, asking us to address and explain the characteristics of black body radiation emission by applying the wave model for radiation. Also, we extend our thanks to Dr. Munawar Karim, Professor of Physics at St. John Fisher College, Rochester, NY for his valuable comments. We would like to extend our gratitude to Mr. Kadakia's colleague, Raymond Mulgrew, Harris RF Communications, in helping him review the text of the book and by providing suggestions to improve the content of this book. Next, we appreciate the constant inspiration from Mr. Kadakia's graduate advisor, Dr. Lyndon Taylor, The University of Texas. Also, we thank Professor Richard A. Matzner, Director Center for Relativity at The University of Texas, for his encouraging remarks for the creation of this book. Further, we thank Dr. Cynthia Galovich from University of Northern California and Professor Walter Toki from Colorado State University for providing useful suggestions on the ideas presented in this book. We thank several medical doctors, in particular Ophthalmologist Dr. Sonal Jadhav and Cardiologist Dr. Sunil Dabhade, for supporting and taking interest in the research.

Mr. Kadakia would like to thank his children for their support and inspiration, and his parents who made the creation of this book possible. Finally, we appreciate the support from readers like you, at various colleges and universities, as well as at various libraries, for providing the support that will lead to the success of this project.

Contents

List of Tables

List of Figures

Prologue

This book is about the true science of light. Many physicists from the beginning of time have failed to recognize the clear understanding of the exact nature of light as it relates to different events, either occurring naturally or created by human efforts. They have portrayed a sense that light behaves as waves in certain events and occasionally it behaves as particles. This ambiguous description of light pondered pioneer physicist Albert Einstein to formulate complex concepts of relativity theory and gravitational theory. Subsequently, renowned physicists, such as Richard Feynman, Roger Penrose, John Wheeler, Charles Misner, Stephen Hawking, and Kip Thorne, extended Einstein's principles to explain the mysteries surrounding distant celestial vast objects, black holes. In their discussion, they and physicist Stephen Hawking have inadvertently stated that black holes have super gravitational fields. From our point of view, light behavior is simple and straight forward in the sense that the rules of Newtonian classical mechanics may be applied without incorporating any special treatment for light. Therefore, black holes absorb light similar to a perfect black body. Further, the behavior of light can be very accurately characterized as a wave, regardless of the type of event. Very close examination of all the experiments performed to prove that light is a particle in those events, could very well be understood if it is modeled as wave.

One of the greatest strengths of this research is that the principles explained in this book provide a rather simplistic view point for several phenomena of complex nature, such as the bending of light caused by refraction as it passes to the medium and time dilation effect experienced by the very high speed moving objects. Our view point extends the ideas suggested by the Portuguese physicist, João Magueijo, a research fellow in theoretical physics at Cambridge University. We have taken one step further, proving his ideas of the Varying Speed of Light (VSL) to be correct. When our concepts are verified, you will gain a clear understanding and explanation of the events related to light, by applying the basic principles of atomic physics described in the book. Here, we have achieved success by taking advantage of the modern techniques and advances made by particle physicists. These physicists formulated the standard model, and the quark extension to the standard model, to describe all of the elements on the periodic table that are found in nature and artificially created. We have provided answers to many questions about the creation of the solar system and the universe that were not answered by previous creation theories, such as the big-bang.

Our theory of relativity, postulated by Mr. Kadakia, designated **Skylativity®**, comes from a simplification of many computations related to sky and the universe, and presents the results with higher accuracy than before. It leads to the formulation of the Universal Unified Field Theory (UUFT), and provides time and space invariant scales for length, time, and mass measures, for every frame of reference. You will discover that Einstein took a risk when he formulated his famous theory of relativity by making unrealistic assumptions. When he stated that time measured in different inertial systems by identical clocks would differ, he ignored the fact that the identical clocks ceased to remain identical in design, when they were stationed in each of the inertial systems. Further, he assumed that the speed of light is a constant in James Maxwell's field equations. More recent advances in technology have verified that the speed of light has varied since the beginning of time and it varies according to frame of reference like an ordinary particle obeying the laws of Newton's mechanics. Our sense of accomplishment and quest will be complete when dedicated physicists and astronomers redirect their resources to promote the ideas of this book and to build a solid foundation for future space expeditions.

1.

Introduction

For more than a century, scientists and engineers have perceived the dual nature of light, a particle and a wave, with great uncertainty. In this book, it is shown that light strictly behaves as a wave. Ever since Albert Einstein devised his theory of relativity [1] to explain the phenomenon of light propagation in free space, very little work has been done to resolve the duality. His subsequent work, to explain the properties of matter at an atomic level and the energy released during a nuclear reaction, popularly described by equation $\mathbf{E} = \mathbf{mc}^2$, heavily relies on the facts that light behaves as a particle and the speed of light is constant **c**. Also, for the computation of important astronomical parameters, such as the distance and velocity of receding and approaching stars and galaxies, and for the age determination of the universe, the Hubble time parameter is utilized by astronomers. The formulation to determine these numbers provides vastly different results, depending on the behavior of the light modeled as a wave or a particle. Further, in this book we have proven that every phenomenon associated with light propagation, and its behavior during the physical and bio-chemical reaction, can be successfully explained by modeling it as a wave. The wave nature and wave model for light, provide the correct physical description of light.

Another question that has puzzled many physicists is whether or not any particle or object can travel at or above the speed of light. Einstein provided an answer by applying the postulates of the special theory of relativity. The theory states that it is impossible to attain the speed of light for any object because its relativistic mass (the sum of rest mass and mass gained because of kinetic energy) rapidly increases toward infinity as the speed of the object approaches the speed of light. We have presented a different view. It is not accurate to state that relativistic mass increases rapidly toward infinity, as the speed of an object approaches the speed of light. The rest mass of an object is a measure of the quantity of matter contained in the object. This rest mass should not depend on the frame of reference. Mass gained by an increase in kinetic energy, is not a real increase in mass. Usually, the gain in kinetic energy of an object, results in a change in the potential energy content of the object. Therefore, within practical limits, it should be possible to accelerate the particles to the speed of light, or even above the speed of light. The upper limit of the speed of light is applicable to energy waves, such as visible light, infrared rays, X-rays, gamma rays, and electromagnetic radiation only, and not to the objects.

Therefore, the principles of the special theory of relativity, the limit on the speed of light **c**, should not be imposed on objects and particles with non-zero rest mass. For example, very tiny particles, such as electrons, are accelerated to the speed of light in a super collider when accelerated under the influence of an electromagnetic field produced by the electric current carrying coils. However, it is possible that when the accelerating electron attains the speed of light, the particle may start to radiate the light, and the kinetic energy of the particle will diminish. Specifically, this is true when the molecules of matter are accelerated to the speed of light because the molecules are comprised of atoms which consist of many protons, neutrons, and orbiting electrons. Thus, a further increase in the speed of any object is prohibited when the atoms and molecules of the object start radiating energy waves. In reality, the speed of the molecules or particles is decreased to conserve energy.

Our assertion is that the speed of a particle is decreased below the speed of light, as the energy content of the particle is dissipated in emission of radiation. This fact is consistent with Erwin Schrödinger's findings in 1925. He derived the wave equations to model the behavior of the light particles as a wave propagating in one, two, and three dimensions. An elementary analysis of the energy radiation by excited atoms may be found in French and Taylor [3]. The analysis, though not rigorous, provides quantitative expressions for irradiative life times and the rate of energy radiation. Further, in the same analysis when an electron orbit transition event occurs, the selection rules for the initial and final quantum states, that govern the possibility of electric dipole transition, are derived.

It is well known that the waves are generated when any particle, string, wire, or metal rod vibrates upon the application of stress. The wave produced is characterized by the frequency of the normal oscillation modes of the system. Schrödinger stated that for every wave, there is a particle motion responsible for its creation. He described that as a wave-particle duality. Clearly, the wave-particle duality is different from the dual nature of the light behavior, wave and particle. Waves at any frequency (sound, electromagnetic, X-rays, laser, maser, and visible light waves), are carriers of energy. No one has described that sound waves consist of particles of matter. Why should a light wave be treated any differently? The only difference between the visible light waves and sound waves is that they differ in frequency. Therefore, the particles causing the radiation of visible light do not constitute a different entity and should not be treated separately. In fact, the light waves in the visible spectrum are released by a process similar to the production of laser and maser. Figure 1.1 illustrates an example of a radiation event caused by the light or thermal energy. Typically, wave energy is released in a radiation event, when an electron in the outer shell makes a transition to a lower shell orbit by excitation. This transition can occur when thermal energy is applied to matter or the high field strength is induced by applying electrical potential, as in the case of solid state laser.

The content of this chapter is organized as follows: In Section 1.1, there is an explanation about why it is important to model the behavior of light very accurately. Then, it is clarified that only the wave model of light is correct. The discussion follows to explain why the absolute speed of light can be attained by particles with non-zero rest mass. In Section 1.2, the example of light trajectory in an accelerating elevator is explained on a wave basis.

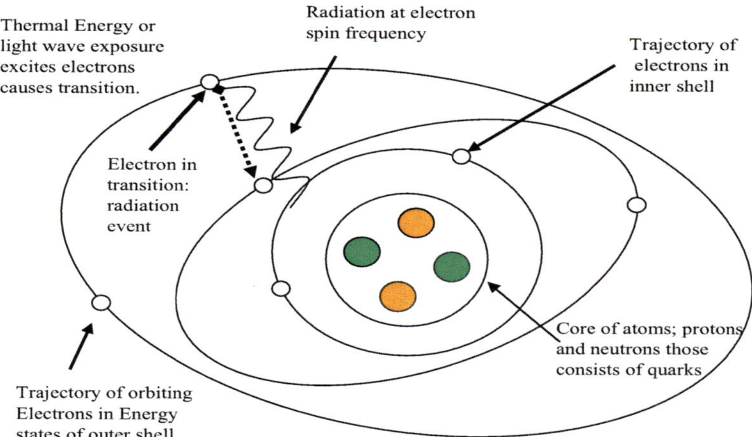

Figure 1.1Radiation events: Orbiting electron in excited state. When orbiting electrons are exposed to the light wave energy, they are excited, triggering an orbital transition event. The excited electrons radiate energy waves at a frequency that corresponds to the energy difference between the initial and final quantum states of an electron in orbit.

In Section 1.3, we will discuss why the bending of light, arriving from distant stars when passing by the sun, is not deflected by the sun's gravity. In Section 1.4, the operation of Michelson's interferometer apparatus is described. Also, it is explained why his experiment does not prove that the classic Lorentz length contraction occurs. In Section 1.4.1, we discuss the assumption that was made, that lead to the wrong conclusion that in turn emphasized the need for the special theory of relativity postulates proposed by Einstein. In Section 1.4.2, an improved design for Michelson's interferometer is suggested. The newly designed apparatus, Sky's interferometer, allows us to precisely measure the orbital speed of the Earth surrounding the sun.

In Section 1.5, we will demonstrate that the results of the experiments of the proponents of particle theory do not provide conclusive evidence for the particle model of light. The results from three experiments, the photoelectric effect, the short wavelength limit X-rays, and the Compton Effect, are investigated in Sections 1.5.1, 1.5.2, and 1.5.3 respectively. Our investigation proves that light is a wave and not a particle, as opposed to the claims made by the performers of the experiments. In Section 1.6, we discuss the details of the single most important effect, Doppler's shift, to prove that light is a wave. Also in this section, we analyze the results of the rotation of blades fan experiment, to prove that light is definitely a wave. Further, zero shifts between the absorption and emission line spectra of various elements, strictly proves that the electromagnetic radiation is a wave. We have included several events associated with light propagation to prove that the light indeed behaves as a wave.

1.1 Light: Wave or Particle?

According to Einstein's special theory of relativity (STR), light exhibits its behavior as a particle and he named the particles photons. Also, in his general theory of relativity (GTR), he stated that a beam of light will bend and slow down under the influence of a force field, such as gravity. In the following two sections, we will show that light does not bend and that its trajectory is not affected by gravity or any other accelerating force. Further, in Section 1.3, we will explain why bending light by the force of gravity would provide catastrophic results when one projects the positions of huge celestial objects nebulae and billions of stars inside neighboring galaxies. In the following section, we will discuss the details of the trajectory of light from a flash light in a moving elevator.

1.2 Accelerating Elevator Experiment

In his example of a man traveling inside a moving elevator, Einstein indicated that a light beam emitted from a flash light carried by the man will strike the wall in front of him at a different spot than at a normal angle to the wall. He claimed that this was due to the bending of light caused by the force of gravity on photons. According to this book, the beam will strike at an offset, because of the upward motion of the elevator. Figure 1.2a illustrates how the path of a light beam is traced when an elevator is stationary with respect to the surface of the Earth. In Figure 1.2b, the path of a light beam, as predicted by Einstein's theory, is shown for the instant the elevator is traveling away from the surface of the Earth. For the same situation, the path projected by new concepts is illustrated in Figure 1,2c.

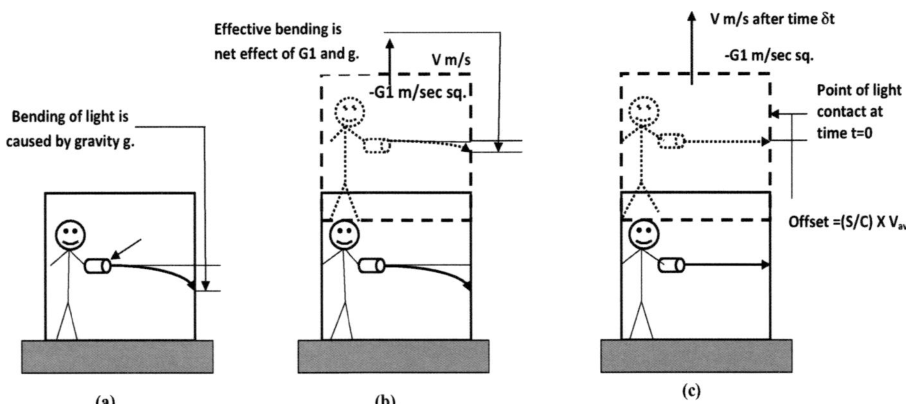

Figure 1.2 Path of light from a flashlight in an elevator. (a) The path of light when the elevator is on the surface of the Earth (Einstein's theory of relativity). (b) The path of light when the elevator is moving upward with an acceleration of **-G1** as projected by Einstein's theory. (c) The path of light when the elevator is moving upward with an acceleration of **-G1** as projected by the **Skylativity®** theory. The point of contact of the light to the wall is always at a right angle (90°) in the case of the **Skylativity®** theory.

One can determine the position of the light beam when it reaches the other side wall of the elevator as follows: the light beam will continue to travel in a straight line after it is separated from the light emitting source. It will take **S/c s** for the light to reach the wall. During that elapsed time, the elevator has moved upward by a distance of **(S/c) × V m** because it is assumed that it is moving upward at the rate of **V m/s**. Here, **S** is the space of the wall from the flashlight and **V** is the velocity of the elevator. Therefore, the observer inside the elevator will visualize that the light ray has bent. Thus, the light will hit the wall at a right angle to:

$$\mathbf{D = (S \div c) \times V} \qquad \textbf{(1.1)}$$

the wall with an offset distance **D** from the point of the normal incident, as if the elevator was stationary. We are implying that the light wave will reach the other side of the wall at a right angle, regardless of the motion of the elevator. Einstein's theory projects the path of the wave will be curved and will bend by the force of gravity. It is very important to compute the distance between the normal position on the wall and the contact point of the light, by the offset method instead of bending by gravity. The main reason is that the universe's gravitational field is constantly changing. All of the celestial objects are moving rapidly. For instance, the tangential velocity of the Earth is at the rate of **29.27 km/s,** orbits surrounding the sun. The tangential velocity of the sun on the ecliptic is at the speed of **216.48 km/s**. Therefore, the effective gravitational force on the light from the distant stars is variable and cannot be computed in real time. The wave model for light and the straight line motion of light is an opportunity that allows mapping the position of the moving celestial object in the sky with very high accuracy than otherwise possible.

A direct effect of the finite velocity of light and the Earth's rotation was discovered by James Bradley in 1927 [7]. He found that the position of all of the stars appears to execute a common annual motion that is evidently a counterpart to the rotation of the Earth around the sun. Therefore, a ray of light coming from a star and striking the objective lens of a telescope appears to come from another direction. Because of the Earth's rotation around the sun, the ray does not reach the eyepiece. Instead, it reaches the wall behind the viewing telescope, as illustrated in Figure 1.3. Some scientists claimed that the measured displacement of the star was partly due to the bending of light caused by the gravitational field from the sun. As explained later, the bending of light is primarily because of the refraction of light caused by the change of refractive indices between the vacuum of free space and air, the Earth's atmosphere. The change of the refractive index between atmospheric air and the vacuum in space should result in a measurable difference in the speed of light. In order to observe a fixed star, the telescope needs to be tilted in the direction of the Earth's rotation that compensates for the displacement in the position of the eyepiece. While the light rays arriving from the star traverse the length **L** of the telescope in time **L/c**, the Earth and the eyepiece of the telescope move by the distance **d** in Figure 1.3. The light rays will strike the eyepiece only if distance:

$$\mathbf{d = v \times (L/c)} \qquad \textbf{(1.2)}$$

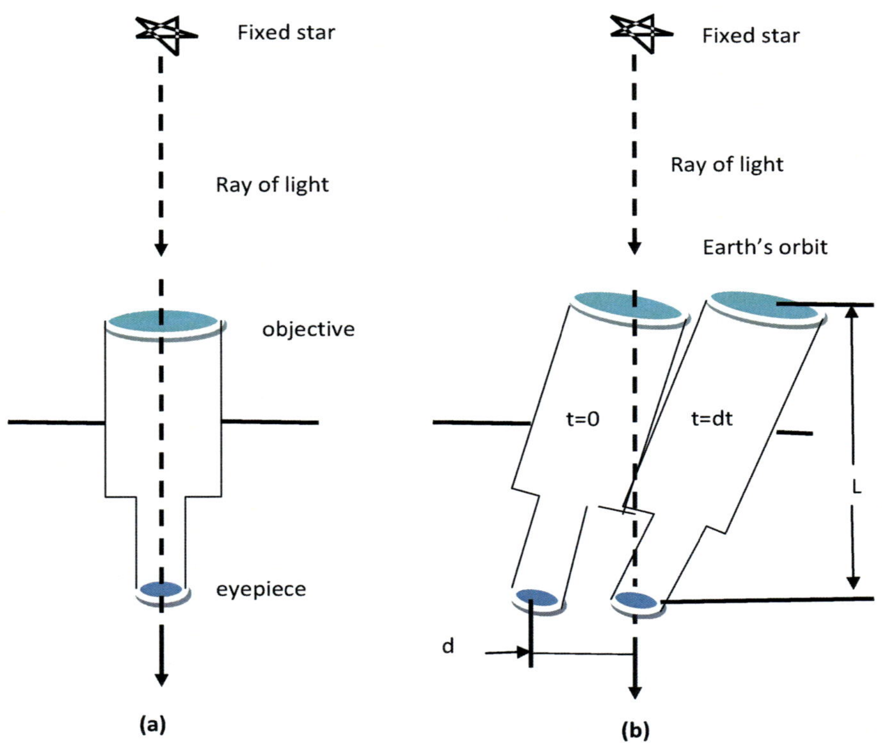

Figure 1.3 Observed position of a star is affected by the Earth's orbital rotation. (a) Observation of a fixed star, the Earth at rest. (b) The telescope is inclined to compensate the Earth's orbital motion. (Courtesy of Max Born, Einstein's theory of Relativity)

The angle of tilt for the telescope is determined by ratio **v/c** where **v** is the speed of the Earth's rotation and **c** is the speed of light which is also equal to ratio **d/L**. The ratio **v/c** is known as the aberration constant **β**. The tilted axial position of a telescope does not point to the true position of the star, but to a point in the sky that is displaced in the direction of **v**. The speed **v** is **30 km/sec**, whereas the speed of light is **~300,000 km/s**. Therefore, the tangent of the angle of tilt is **1/10,000**. The value of **β** can be precisely determined by measuring the elliptical motion of a fixed star, such as a polestar. This measurement provides a very accurate technique for the computation of the speed of light **c**.

According to the theory of relativity, the material mass of a light photon is zero. Therefore, by definition, a photon cannot be a particle, because particles have a finite mass in any frame of reference and a fixed position for the center of gravity. In fact, in physics, the mass of an object is defined as a measure of the quantity of the matter contained. Since a photon does not contain any matter, it has zero rest mass. In this book, we support the viewpoint that a light beam always travels in space as a wave and in a straight line. It will not bend or deflect, and need not obey Sir Isaac Newton's gravitational laws, because gravity does not have any effect on an object with zero rest mass. On the other hand, it is important to note that there is no event in which the gravity effect from the light waves on other waves or particles is observed. Meaning, no matter how strong the light wave energy is projected from any source, no force of gravity from the light waves is detected. This fact provides further evidence and increases confidence in our proposed theory. It is well known that the force of gravity on an object on the Earth with mass **m kg** is given by:

$$\mathbf{F = ma = mg} \qquad \textbf{(1.3)}$$

from Newton's Second Law

Einstein proposed that a photon has non-zero mass due to inertia. From his formula, $\mathbf{E=mc^2}$, the mass equivalent of **E units** of light energy is:

$$\mathbf{m = E/(Square\ of\ the\ Speed\ of\ Light) = E/c^2} \qquad \textbf{(1.4)}$$

We like to name this energy mass of photons as virtual mass because it does not exist physically and it is not real. Also, when it relates to photons, the energy mass is not concentrated. It is distributed with respect to its position. Therefore, the center of gravity for a photon is not known. Furthermore, as stated by the Heisenberg uncertainty principle, one cannot determine the position and momentum of any particle at the same time with the same precision [3]. Our point is that the photon and light waves do not have a center of gravity, and they have zero rest mass. Therefore, the position of center of gravity for light waves is indeterminate. The gravitational force acts between the two bodies with a deterministic center of mass. Hence, a light wave should not bend under the influence of the gravitational field. In the next section, we will further describe the consequence of the bending of light by the force of gravity.

1.3 Deflection of Light from the Star by the Sun

According to GTR, in the presence of gravitational fields, the velocity of material bodies or of light can assume any numeric value. In an arbitrary Gaussian coordinate system, not only does the velocity of light become different, but the light rays no longer remain straight [7]. We agree that the speed of light varies. We disagree that the light rays will bend. Einstein tried to explain the bending of light rays from distant stars because of the gravity of the sun. Two scientists attempted to verify Einstein's prediction of the deflection of a ray of light coming from a star by **1.75 s**. One was German mathematician, Johann Georg von Soldner in 1801, and the other was British astronomer, Arthur Stanley Eddington as late as May 29, 1919. Two British expeditions were sent to observe a total eclipse of the sun, one to the west coast of Africa, the other to the north of Brazil. They returned with a number of photographs of the stars surrounding the sun. The results obtained from the photographs were announced on November 6, 1919. The displacement from the pictures was **1.75 arc s** which was in close agreement with Einstein's prediction. Many scientists applied the results of this test to conclude that the principles of Einstein's theory of relativity were correct. Figure 1.4a illustrates the situation in which a ray of light arriving from a fixed star passes close to the sun and is observed on the Earth. The ray will be attracted toward the sun by the force of gravity from the sun, and will create a somewhat concave trajectory, as indicated in Figure 1.4a.

From our point of view, the light waves follow a curved path for a different reason. The light rays arriving from the distant stars are deflected as they pass by the sun because they are refracted by the hydrogen gas atmosphere of the sun. Also, the light rays from the stars are further refracted by the Earth's atmosphere. The deflection of light by gravity effects could produce disastrous results when they suffer bending, due to multiple stars. One such incident is depicted in Figure 1.4b. It is clear that the gravitational bending of light by multiple stars will not permit mapping the sky with billions of stars in the universe with any acceptable level of accuracy. Now, let us compare the bending of a light ray, arriving from a distant star, because of the change in the refractive index of the atmosphere from a vacuum (free space) and the bending due to the sun's gravity. In the following, we will apply Newton's laws to compute the bending of light by the gravitational field. Also, we will apply Snell's law, named for Willebrord Snellius, to compute the deflection of light as a result of the passage of light through the Earth's atmosphere.

Computation – Refraction of light by Snell's law vs. the bending of light by gravity:

A. Bending of light in the proximity of the sun because of the sun's gravity:

mass of the sun $\mathbf{M_s = 1.98 \times 10^{30}}$ **kg,**
mass of the Earth $\mathbf{M_e = 5.98 \times 10^{24}}$ **kg**
radius of the sun $\mathbf{= 6.95 \times 10^{8}}$ **m**
radius of the Earth $\mathbf{= 6.38 \times 10^{6}}$ **m**

universal constant $\mathbf{G = 6.6726 \times 10^{-11}\ m^3/kg\text{-}s^2}$
average radius for the Earth's orbit $\mathbf{= 1.49 \times 10^{11}\ m}$

the computed deflection because of the gravitational bending of light,
due to the sun's gravity, using Newton's Kinematics: $\mathbf{D_b = (0.5)\ G_{sp} \times t^2}$

to calculate $\mathbf{G_{sp}}$:

$$\mathbf{G_{sp} = G \times M_s / (R_{oa})^2}$$
$$\mathbf{= 6.6726 \times 10^{-11} \times 1.98 \times 10^{30} / (1.49 \times 10^{11})^2}$$
$$\mathbf{= 5.951 \times 10^{-3}\ m/s^2}$$

$$\mathbf{G_{ss} = 273.52\ m/s^2}$$

$$\mathbf{D_b = (0.5)(5.951 \times 10^{-3}) \times (8 \times 60)^2}$$
$$\mathbf{= 685.55\ m}$$

B. Bending of light as it travels through the Earth's atmosphere using Snell's Law:

Refractive index of vacuum: $\mu_v = \mathbf{1.0}$
and that for atmosphere air: $\mu_a = \mathbf{1.003}$
notice that μ_a could vary a lot depending on the density of the air
let ϕ_i angle of the incident ray at the vacuum and air interface
ϕ_r angle of the refracted ray received by the observing telescope

from Snell's law: $\mu_v \mathbf{Sin}\phi_i = \mu_a \mathbf{Sin}\phi_r$ **(1.5)**

for a normal ray: $\phi_i = \mathbf{90^o}$
substituting into the equation (1.5)
gives $\mathbf{Sin}\phi_r = \mathbf{0.997}$ and $\phi_r = \mathbf{85.5607}$

the deflection ϕ_d is: **90 – 85.56 = 4.4393° = 0.07749** radians

the maximum deflection:

$$
\begin{aligned}
\mathbf{d \tan \phi_d} &= \mathbf{8 \times 60 \times 3 \times 10^5 \times \tan (4.4393)} \\
&= \mathbf{144 \times 7.7636 \times 10^4\ km} \\
&= \mathbf{1117.96 \times 10^4\ km} \\
&= \mathbf{11.1796 \times 10^6\ km}
\end{aligned}
$$

From these computations, it is obvious that the deflection by Snell's law is > **16.3M** times the deflection due to the gravitation bending of light from the sun's gravity. The effective bending of light by gravity is a lot smaller than the accuracy of measurement.

The results of the computations indicate that the bending due to the refractive index change using Snell's law is higher by six levels of magnitude than the deflection of the light due to the gravitational pull of the sun. Figure 1.4b illustrates that a light arriving from a distant star to the Earth may experience multiple bending effects as it passes by several stars. This is true if we assume that the force of gravity from the different stars will bend the light waves several times. It is obvious why such deflection of light from other stars, before it is observed on the Earth, will make the task of mapping the sky from the Earth extremely difficult. The situation is described when the light ray, arriving from a distant star, passes by a star closer to the sun and also passes by the sun before it is observed on the Earth. The multiple bending effects on the light, by thousands of stars with differences in distance of thousands of light years, will make the task of locating the stars in the sky impossible because we know that all celestial objects are in constant motion in relation to each other. This makes the matter worse. In addition to that, the light waves arriving from the different stellar objects, the distant stars and galaxies, portray different time slices and delays. Thus, it is essential that the sky be mapped in more than one universe. We will address this issue in later chapters. Therefore, the photographs of the stars during a total solar eclipse should be taken outside of the Earth's atmosphere. The fact that the deflection value is the same within the accuracy of measurements at two different locations, Africa and Brazil, could possibly lead to an incorrect conclusion that light bends and is affected by the gravity of the sun because the correction factor to compensate for the bending caused by the refractive index, varies with time and location.

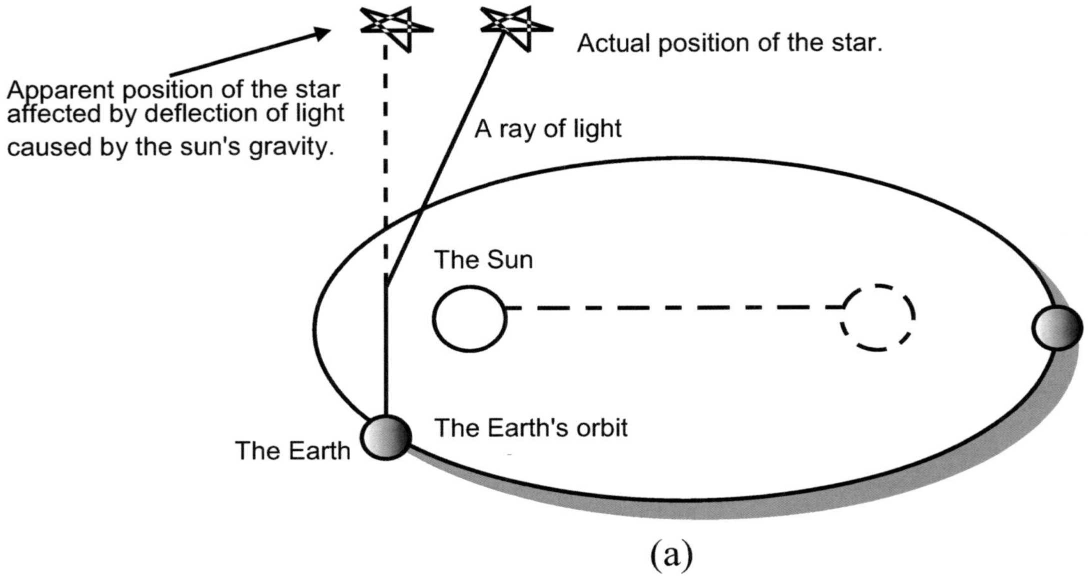

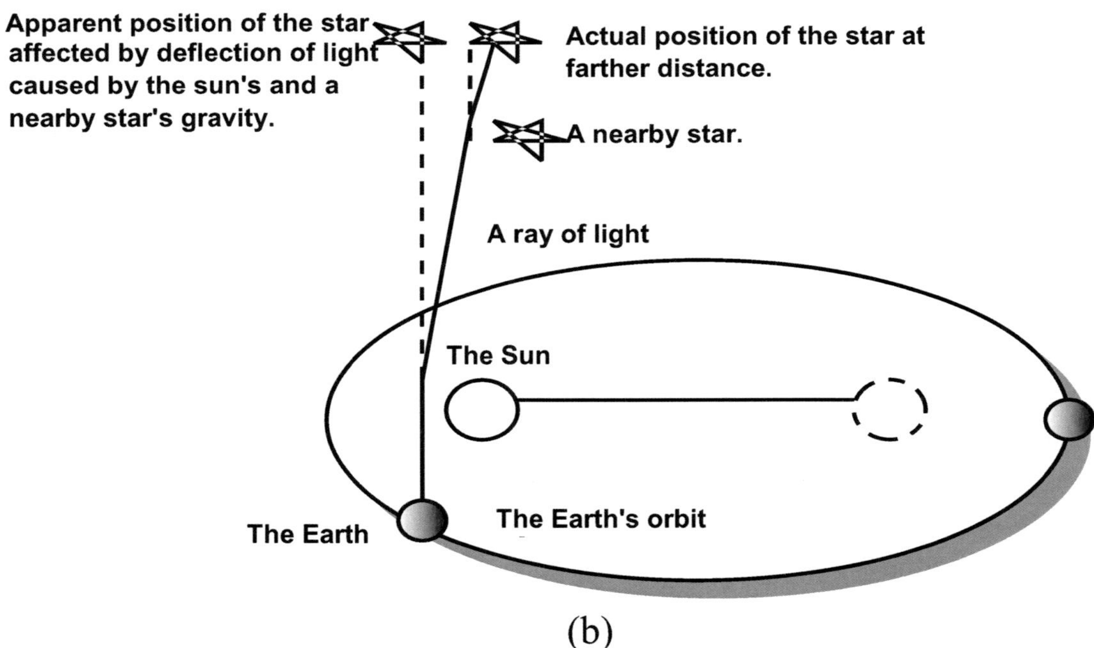

Figure 1.4 Position of the star is affected by the bending of light. (a) The deflection of a ray of light from a star by the sun. (b) The trouble in computations: the bending of light is caused by multiple stars.

Let us suppose that a photon is a particle with inertial mass, as proposed by Einstein. The light beams bend toward the Earth as they experience the force of gravity, as in the elevator experiment. He has stated that the photons cannot be accelerated or retarded, and maintain the constant speed of **c** in space. Since the light beam has bent, the specific photons' velocity vector has changed. This is a contradiction, which Einstein's theory cannot explain. He stated that the speed of the celestial stars changes the frequency of the photons arriving from the distant stars. The shift in frequency is toward the blue end of the spectrum for photons from stars approaching closer to the Earth and is toward the red end of the visible spectrum for receding photons from receding stars. Interestingly enough, he mentioned the frequency, which is a property associated with the waves, not the particles. Furthermore, the trajectory of light arriving from any star of another galaxy will suffer multiple bending if it will bend by force of gravity of stars because it passes by many stars. The light from the stellar object will be wandering in the universe and would not reach the eyepiece of the telescopes or sensors on the Earth. The wandering light rays will complicate the computation of the distance of a star, based on the reception of the light arriving from the star, to such an extent that the results will produce an unacceptable number of errors. From this explanation of the various phenomena and computations, it is evident that the light arriving from distant stars does not bend by force of suns' gravity. Further it allows us to conclude, the light behaves strictly as waves and its behavior need not be confused as a particle.

In the next section, we will describe the details of Michelson's interferometer. The unusual results from his experiments provided fuel for Einstein to formulate his postulates of STR and GTR. Subsequently, we will show that the incorrect observations made from the outcome of the interferometer experiments, lead to wrong conclusions. Therefore, we should apply the wave nature philosophy for conducting advanced research in the study of light waves, rather than following the principles of Einstein and others describing the ambiguous light behavior, sometimes as a particle and occasionally behaving as waves. One of the purposes of Michelson's interferometer was to measure speed of earth's orbital motion around the sun. To perform this measurement, he decided to compare light path time delay differences which were orthogonal to each other. To determine path delays he was interested in observing the differences in radii of O-ring patterns created by interference effects of light waves reflected from three mirrors and a source of light. He monitored light arriving from different paths through an eyepiece of a telescope.

1.4 Michelson's Interferometer

To measure the speed of the Earth's orbit surrounding the sun, Albert Abraham Michelson and Edward Morley performed their interferometer experiment. For reasons explained later, their measurement did not lead them to estimate the speed of the Earth. Michelson's apparatus consists of the radiation source of laser (monochromatic light) of a single wavelength, three mirrors **A**, **B** and **C**, and a means of detection, a telescope for observing the reflected light. Figure 1.5 illustrates a simplified diagram of the arrangement of mirrors, a laser light source, and a telescope.

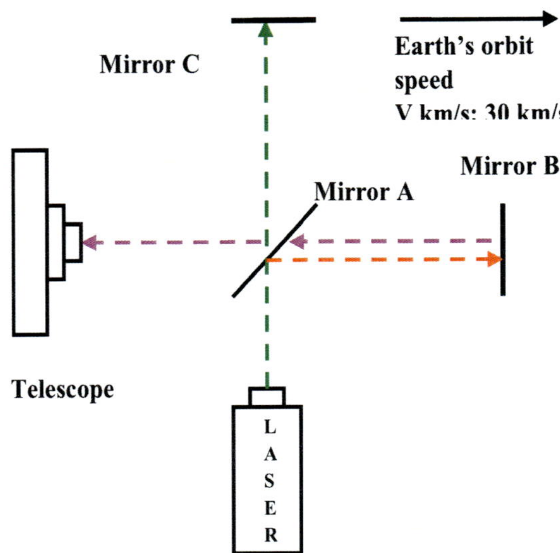

Figure 1.5 Michelson's interferometer. In his experiment, the difference in delays between two possible light paths may result in an interference pattern to be observed by a telescope.

Classically, the light path lengths **ABA** for mirrors **A** and **B**, and **ACA** for mirrors **A** and **C,** should differ because mirror **B** and the entire apparatus is moving in the direction of the Earth's orbit surrounding the sun. The position of mirror **C** is normal for the Earth's speed and should move in accordance with mirror **A**. Also, though the Earth rotates on its axis, its rotation should not affect the relative positions of mirrors **A**, **B** and **C**. If the entire apparatus were stationary, that is, if the Earth did not have orbital motion, then the path lengths **ABA** and **ACA** would be equal to **2L**, **L** is the distance between mirrors **A** and **B** or **A** and **C**. Notice that the distances **AB** and **BA** are traveled at different speeds by the light caused by the Earth's orbital speed, and the switched direction between the forward and reflected path. Therefore, the light waves received at **A**, from mirrors **B** and **C**, should be different in time and phase, when observed by the telescope. The measurement of the time difference and the determination of the speed of light, by this experiment, was one of the primary purposes of the experiment. As we will see, Michelson was unable to achieve the desired result for the following reasons.

Michelson expected that the time difference would show a fringe pattern that would reflect the speed of the Earth; instead he saw no shift and null result [2]. Now, we know that the speed of the Earth's orbit surrounding the sun is **30 km/s**. An excellent analysis of the events which occur in this apparatus may be found in [2]. However, the facts illustrated and the details of the experiment, as described by Robert Mills in [2], are not accurate. To calculate the time, he is using the relative speed of light **c - v** in the forward direction and **c + v** in the reflected direction. To compute the length of the paths, he is applying the constant speed of light **c** in the formulae. This computation is inconsistent. If we change the speed of light to **c**, by applying the Galileo transformation, and apply it consistently, the sum of the forward and reflected path length will always be **2L**, regardless of the orbit speed. This is the main reason that he received a null result every time. It is important to note that the speed of light is constant **c** in a single medium, as long as the source and destination (telescope) are stationary with respect to the frame of reference. In our experiment, the entire apparatus and the Earth (frame of reference) move with the orbit speed. Therefore, the speed of light is constant in the forward and reflected path. According to this assumption, the length of path:

in a forward direction **Lpf** = **mirror distance AB + distance moved by B in time for the light to reach AB**

time for light to reach distance **L** = **L/c**

forward direction path length L_{pf} = $L + (L \times V)/c = L(c + V)/c$

reflected path length L_{pr} = **mirror distance BA – distance moved by A in time for the light to reach BA**

$$= L - (L \times V)/c$$
$$= L(c - V)/c$$

when we add paths

$$= L_{pf} + L_{pr}$$
$$= L(c + V)/c + L(c - V)/c$$
$$= 2L$$

Naturally, when the total path lengths **ABA** and **ACA** are equal, there is no phase difference detected by the telescope at all times. This proves that the analysis described by Mills is not correct. In the formulae, we have made an assumption that the path length **AB** is always **L**. This is a correct assumption, because on Earth we cannot design any apparatus which would be stationary, with respect to the sun. Also, we will explain in the next paragraph, why the distances will not change, which is caused by a contraction in the length of the mirrors **A** and **B**. The contraction computed by the Lorentz equation, is a virtual contraction and is not a real event.

Einstein attempted to explain the null results for the fringe pattern in Michelson's experiment as follows: He stated that the speed of light along the direction of the Earth's orbit, and in the opposite direction for a reflected ray of light, has the same value **c** as a constant. Classically, the speed of light in the forward path will be c – **V** and in the reflected path will be **c + V**. In order to nullify the classical difference in the path lengths and time, he proposed that time will be slower in the forward path than the classical time. Therefore, the path length seen by the light will be contracted to value **L**. The reflected path, according to the classical procedure, the path length will be shorter than length **L**. In order to compensate for the decrease in length, the relativistic length determination should expand the length to **L**. For that, Einstein's postulate should require that time speed-up for the return path of the light. Einstein clearly stated that time is always slower for all other inertial systems, as compared to the preferred frame of reference. Therefore, his postulates fail to explain the null result of the experiment. In Sub-section 1.4.1, we will describe what causes provided the wrong results in Michelson's interferometer experiment.

1.4.1 Analysis of Path Lengths and Delays

Now, let us examine and discover why Michelson's conclusion, which implies that the relativistic contraction of path **ABA**, to compensate for the effective increase in the path length, is incorrect. Figures 1.6 and 1.7 illustrate the path length in the direction of the motion of the Earth and the reflected path. We agree that the forward path traversed, is larger than length **L** and the reflected path is smaller than **L**. Also, if we assume that the speed of light changes according to the relative speed of light, then the total path length **ABA**, will be longer than **ACA**. In that case, the analysis in [2] the path length **ABA**=$2Lc^2/c^2$-V^2, is correct. In the description, the path length **ACA** is $2Lc/\ (c^2-V^2)^{1/2}$ **[2]**. We agree that the value of the path length **ABA** is right. The value of the path length **ACA** is **2L** and not the value of the expression. According to our discussion, **ACA** should be exactly **2L** long because the light is a wave and travels in a normal direction to the surface of mirror **C**, unaffected by the Earth's orbital motion. In essence, the path lengths **ABA** and **ACA** are different, and the difference should be detected as the speed of the Earth.

Let us suppose that the distance **L** in an apparatus is **1 m**. Also, for illustration purposes, we assume that the wavelength of the laser light used is **400 nm**. Then, the difference between the paths **ABA** and **ACA** in our discussion will be:

$$L_{diff.} = \text{path length ABA} - \text{path length ACA}$$

$$= (2Lc^2/\ (c^2-V^2))\ -\ 2L$$
$$= 2LV^2/\ (c^2-V^2)$$

$$L_{diff.} = 20.0277012\ \text{nm}$$

for **c = 299792458 m/s** and
V = 30000 m/s.

The path difference L_{diff} corresponds to a phase shift of **18.02493108°**. This phase shift should be detected by the telescope. Michelson claimed that his apparatus was capable of detecting the phase shift and the length of the difference of **1/100** of the wavelength, that is, a distance of **4 nm** in our example of the laser light source. To create a fringe pattern, he will have to add a length offset of **90 nm**, that is, **4** to **5** times the distance he wants to measure. To add a small distance of **90 nm** in **1 m** length, with any mechanical means would be a formidable task. Especially, if he is achieving this by tilting mirrors, etc. Therefore, he failed to see any fringes without adjusting the mirrors.

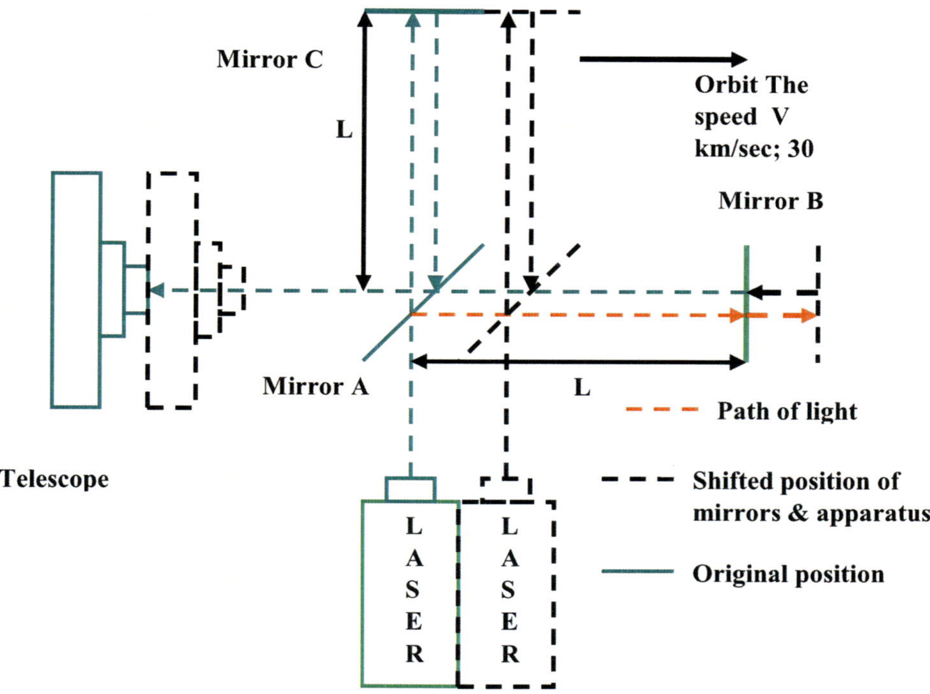

Figure 1.6 Path of light in the direction of the Earth's orbit. The path of light is longer when mirror **B** is shifted in the direction of the Earth's orbital motion surrounding the sun as opposed to the stationary Earth.

Further, this path difference should result in an apparent shift in the frequency of the received signal, as suggested by Christian Johann Doppler. In fact, the principle of Doppler's shift is regularly applied to determine the speed of the celestial objects. The Doppler shift is also applied in the radar computation to determine the location and speed of the mobile target for military applications. However, the Doppler shift can only occur if there is a motion between the source and the observer. In this experiment, the source of the laser light, the telescope observer, and all mirrors, is moving together with the Earth. Therefore, the Doppler shift is not observed.

According to his analysis, Michelson specified that the path difference and the time delay between **ABA** and **ACA** would be cancelled by the effective decrease in the path length **ABA,** caused by the relativistic contraction. Obviously, his thought was incorrect. If the contraction occurs in two mirrors, the contraction value can't exactly match the difference in the length of the contraction that corresponds to the Lorentz transformation, and was in accordance with Einstein's theory of relativity. In his apparatus, the majority of the path traversed by the light is an empty space. It does not make sense to say that the space is contracted. He states that the mirror **B** experienced contraction because of the Earth's motion. Such a contraction is not realistic for the speed **30 km/s**, and the contraction of mirrors **A** and **B** will result in an increase of the path length, and not a decrease between the forward path **AB** and the reflected path **BA**.

There are many reasons for which the contraction can't occur in reality. For one, the strong and weak forces in the atom structure play a predominant role for the determination of the volume or size of objects made out of matter. Secondly, the force of gravity from the large celestial objects, such as the Earth, has a very small effect on the fundamental particles, protons, neutrons, and electrons that form the atoms. If the force of the gravity, compared to the force of the charge in the nucleus particles, is as small as $\mathbf{10^{-42}}$ [7], the effect of the change of momentum, due to the orbital speed of atoms, cannot be big enough to result in contraction. Specifically, contraction can't be deterministic and accurate to a value precisely predicted by the Lorentz transformation. Also, we know that one needs to supply or remove a lot of the thermal energy to a glass or metal surface to cause an expansion or contraction of this proportion. Further, the different path length **AB** in the forward path and **BA** of the reflected path is not the characteristic of the light signal. The path and the time taken to traverse both paths would be different regardless of what type of signal and sensing mechanism is employed. In addition, the apparatus and the mirrors are under the influence of the motion of the Earth around its own axis. The axial rotation speed of the Earth at the equator is **0.463 km/s**, which is smaller by a factor of **64.66** than its orbit speed surrounding the sun. When Michelson was looking for fringes caused by the phase shift between the signals from paths **ABA** and **ACA**, the Earth's axial rotation can affect his results significantly. Michelson observed the fringe pattern when he adjusted the tilt angle of mirror **A**. The reason for this result was that he inadvertently changed the length of arm **AB** that caused the phase difference in the signals received by the telescope, from mirror **B**, path **AB**, and mirror **C**, path **AC**.

After several hours of discussion with Dr. Amin Jaffer, Engineering Fellow at Raytheon Company, Space and Airborne Systems, we concluded that the setup of Michelson's interferometer is not the correct method to measure the speed of the Earth's orbital motion surrounding the sun, because it assumes that the reflection of the light from mirror **B** occurs in zero time. As far as we know, the analysis of this experiment by all physicists to date has not included any correction factor or the effect of the time of reflection in the computation. In the next section, we are proposing an improvement to Michelson's interferometer. Our proposed design eliminates the need for multiple reflections of light from mirrors **B** and **C**. Instead, we are relying on one time event. The light from mirror **A** is sensed by photo sensors. The onset of the change of potential detected by the photo sensors is measured by an oscilloscope. The light rays from the laser source are switched on-off periodically, to create a repeated event at the sensors of the light. In this arrangement, the mirrors **B** and **C** in Michelson's apparatus are replaced by the photo sensors **B** and **C**. These sensors do not reflect light, an arrangement that is different than described in the previous experiment. As we will discover, this technique does not suffer the disadvantages of the prior experiments. One need not take into account the time when the light beam is reflected and reverses its direction from mirrors **B** and **C**. Also, in this method, the classical paradox of the different speeds of light in different directions is eliminated. As we will see, this improved arrangement lacks all of the drawbacks of the older apparatus.

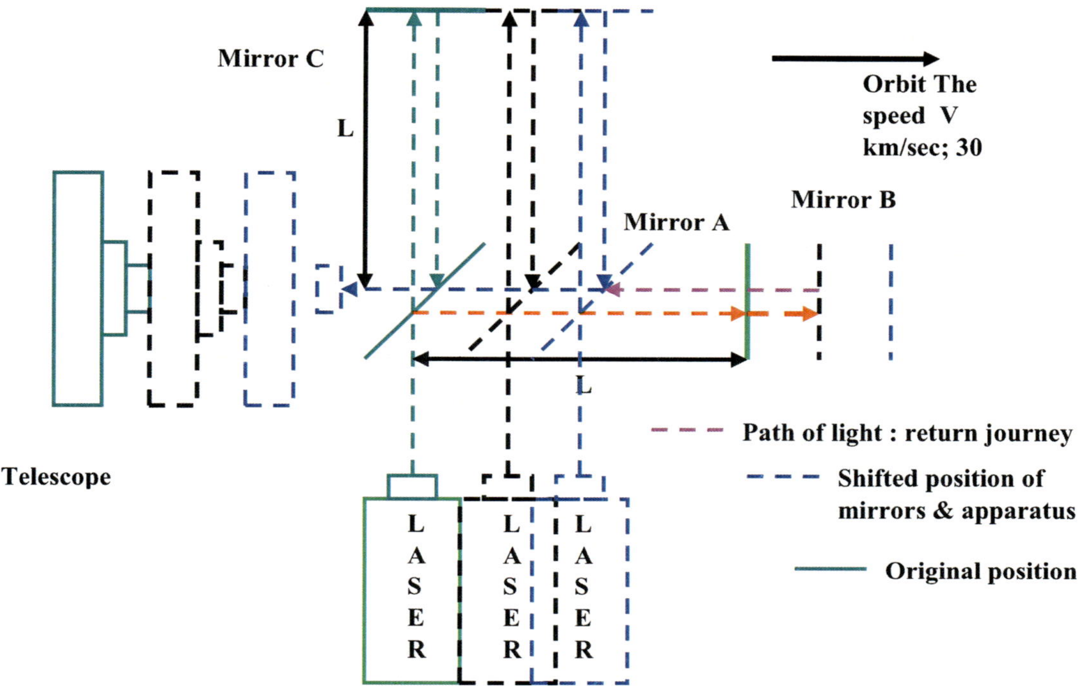

Figure 1.7 Michelson's Interferometer: Reflected Light Path. The path of light is shorter when it is reflected as it travels in the opposite direction of the Earth's orbital motion surrounding the sun.

1.4.2 Sky's Improved Interferometer

In the section 1.4.1, we explained what went wrong in Michelson's interferometer experiment that prevented the successful measurement of the speed of the Earth's axial rotation, by capturing the time of the roundtrip for light from a reflected mirror arrangement. The main drawback of the interferometer was that it relied on a fact the reflection of the light wave from the mirror occurs in zero time. Also, the apparatus is incapable of discerning the observed time differential between two paths, one normal to the axis of the Earth's rotation and the other in the direction of rotation. The pattern of the interference of light from both paths is not formed by one wave. The **O** ring interference is observed as a result of the time difference between two paths and is read in a period of several milliseconds. As explained in the section 1.4.1, the time difference in terms of the wave period, because of the Earth's rotation, is only **18°** of the wave period for the visible light source. Therefore, the time of the reflection from the mirrors should not be neglected.

Now that we have addressed the pitfalls of Michelson's interferometer, we will suggest an improved design for the interferometer and we will designate it as Sky's interferometer. Before we describe the new design and principles, let us propose an alternate method to measure the speed of the Earth's rotation. One way is to use the source of the energy waves that travel in the direction of the Earth's rotation at a speed much smaller than the speed of light. The speed of these waves will be nearing the speed of the Earth's rotation, which is **30 km/s**. One drawback of this method is that it requires that one have control over the speed of the energy waves. Controlling speed of an energy wave is not a simple concept to comprehend.

Therefore, we are proposing a different solution. Our instrument relies on the principle of time measurement of non-reflected waves. The time difference is measured between a normal wave and a wave in the direction of the Earth's orbit, in a onetime event. We will assume that the orbital speed of the Earth's rotation surround the sun is constant over the time frame of measurement. This is a very safe and accurate assumption because the orbital period of the Earth's rotation has not changed to a great extent over the period of a century. The time of measurement of this experiment is short, for instance, a few minutes. Three events synchronized with a common trigger time-base are employed. The arrangement is similar to Michelson's interferometer. First a time-base is used to drive a source of light or laser on the left side of a transparent glass lenses **A**. The time delay of light sensed by two identical semiconductor photo electron sensors is measured by two channels on an oscilloscope. One of the sensors is located at a distance of one meter in the direction of the rotation of the Earth's orbit. The other sensor is located in a direction normal to the rotation of the Earth's orbit at the same distance. The time-bases of the oscilloscope, that measure the response from two sensors, is swept by the time-base that drives the laser light source.

The arrangement of our apparatus is demonstrated in Figure 1.8. The apparatus consists of a laser light source **G**, a transparent mirror **A**, a photo sensor **B**, and an identical sensor **C**. The sensor **B** is located above **A** at a distance of **1 m** and sensor **C** is located on the right side at the same distance of **1 m** from **A**. The signals from both sensors are received at the input terminals of two channels of an oscilloscope, with identical probes and input characteristics. What we are looking for, is the onset of the deflection of the scope beam by the voltage signal response from the sensors. Since the trigger time-base and the component distances, and all characteristics of measurement, can be most accurately tailored to make it identical in all aspects, the differential time-delay measurement between sensor **B** and **C,** truly reflects the rotation speed of the Earth. To improve the accuracy of measurement, both channels of the scope should be run in a single shot trigger event monitor mode.

Let time **t** for the signal to propagate from the laser to sensor **C**. Also, **δt** is the time difference between the onset of the transition of the signal from sensor **B** and sensor **C**. Then the Earth's orbital speed can be computed as:

$$\mathbf{V = c \times \delta t / t} \qquad \mathbf{(1.6)}$$

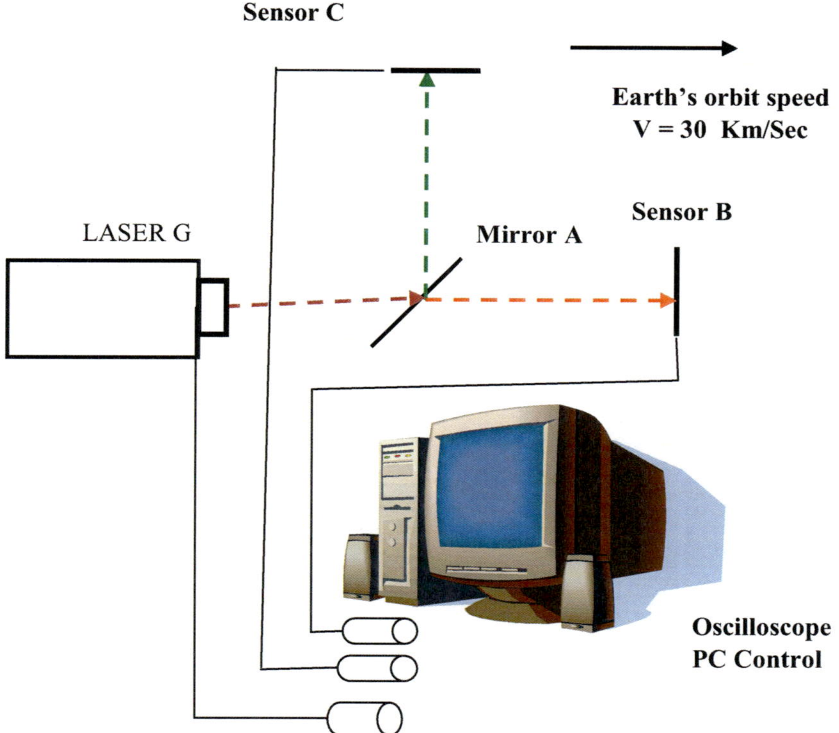

Figure 1.8 Sky's Apparatus: Single Shot Interferometer. The method suggested by us relies on the fact that the path delays are measured in one event without reflections from mirrors.

In section 1.5, we will describe the details of the various experiments performed to validate that light behaves as a particle in certain events. Then, we will discuss the results of those experiments performed, that intended to prove the postulates of the theory of relativity from Einstein. Next, we will demonstrate that the results of those experiments do not provide conclusive evidence and their claims are in error regarding the nature of light as a particle in those instances.

1.5 Advocates of Particle Theory and Experiments

The proponents, including Einstein, of the theory that light consists of photon particles, used the results from several experiments to validate the concept. We will discuss three experiments. Their evidence indicates that the energy possessed by the individual photon of frequency **ν** is expressed by the relation:

$$\mathbf{E_{photon} = h \times \nu} \quad \textbf{(1.7)}$$

here **h** is known as Planck's constant

In particular, we will investigate the details of the following the experiments in great depth, in order to correctly understand the behavior of light waves:

1. The Photoelectric effect
2. Short wavelength limit X- rays
3. The Compton effect

We discovered that the evidence from those experiments does not necessarily prove that light consists of photon particles. Our observation is based on the fact that the outcome of the experiment would not be different, even for a wave model for light. Let us look at the details of each experiment and explain why their results do not conclude that the light is a particle and instead is a wave.

1.5.1 The Photoelectric Effect

In this section, we will discuss the details of the experiment for photoelectric effect. It is well known, that free electrons are ejected from a clean metal surface when exposed to light energy. The electrons are called photoelectrons and the ejection by light is known as the photoelectric effect. It is determined, that for a given value of frequency $\boldsymbol{\nu}$ or wavelength $\boldsymbol{\lambda}$ for incident light, there is a spread of photoelectron energies down to zero. However, the maximum kinetic energy $\mathbf{K_{max}}$ of photoelectrons is sharply defined and varies linearly with $\boldsymbol{\nu}$. The energy $\mathbf{K_{max}}$ does not depend on the intensity of light, but only on its frequency. Experimentally, it is verified that the high intensity of the incident light results in increased numbers of photoelectrons ejection, but not more energy per electron. Further, below a minimum threshold frequency $\boldsymbol{\nu_0}$ for the incident light, no photoelectrons are ejected, regardless of the intensity of the incident light. The threshold frequency $\boldsymbol{\nu_0}$, is the characteristic of the metal being used as the photo emitter.

Figure 1.9 illustrates the photoelectron ejection by colliding the incident light. A closer look at the ejection of photoelectrons, due to the incident light, indicates a conclusion that is contrary to Robert A. Milikan's observation. In Table 1.1, the maximum kinetic energy of photoelectrons vs. the frequency of incident light rays is summarized, and the data is plotted in Figure 1.10. We provide the following explanation for our conclusion, in favor of the wave nature for the energy that is incident to eject electrons from the orbits. The data was reproduced, courtesy of Milikan, phys. Rev. 7, 355 (1916).

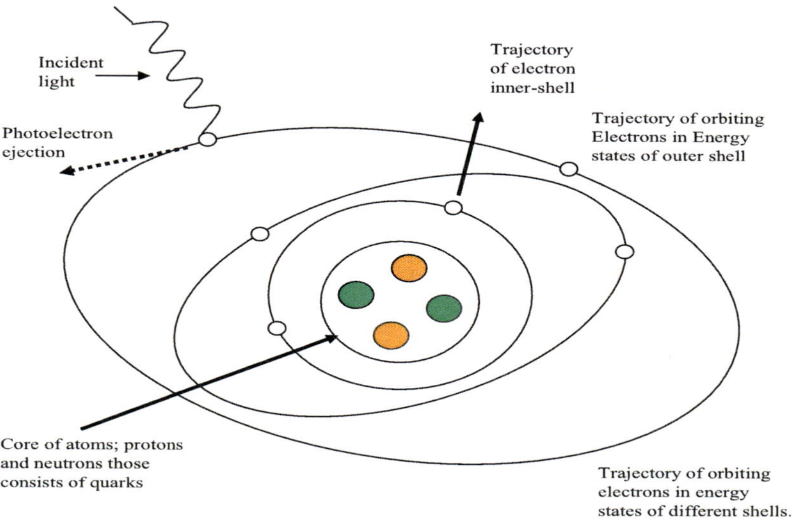

Figure 1.9 Photoelectron emissions by incident light on a clean metal surface. When bound electrons in metal are exposed to light, they absorb quantum of energy and escape for conduction of current.

The energy $\mathbf{K_{max}}$ depends on the frequency and not the intensity of the incident light, because the ejected electron from a specific orbit absorbs the light energy in a quantized maximum amount at the frequency of incident light. The quantity of energy absorbed by an orbiting electron is a function of the de Broglie wavelength representing the electron in orbit. At the frequencies of the light incident corresponding at or above the de Broglie wavelength, the electron may absorb more energy which is required to alter the quantum state of the electron. At the higher frequency, light waves penetrate the inner shells and remove the electrons from the interior orbits, after those electrons absorb the light energy. Obviously, the lower frequency of light with low energy can't penetrate electrons in inner orbits. The kinetic energy of electrons from the inner orbits, is higher than the electrons in the outer orbits, because the orbital speed of electrons in the interior orbits is higher than the electrons in exterior orbits. The fact that the incident light has a frequency below threshold $\mathbf{\nu_0}$, does not result in any ejection, regardless of the intensity, which proves that below that frequency, the light energy is not absorbed by electrons in any orbit. Therefore, the results of this experiment do not prove that light consists of photon particles.

Also, it is determined that the photoelectric current is directly proportional to the intensity of the incident light. The higher the intensity of light, the more photoelectrons are ejected if the frequency of light is above the threshold frequency. However, this linear relationship does not prove that light is a particle. More electrons on the metal may absorb energy from the light waves, if the intensity is high. Thus, light waves with high intensity, may eject more photoelectrons and increase the conduction. It is surprising that Einstein's photoelectric effect equation works equally well if the light is modeled as a wave or a particle. The photoelectron may be emitted if an electron on a metal anode absorbs a quantum of energy, whether or not the light was modeled as a particle or wave.

Table 1.1 Frequency of light waves vs. Maximum KE of photoelectrons

Frequency 10×THz	Kinetic Energy Electron volts
48.5	-2.33
55	-2.03
69	-1.5
74	-1.3
82	-0.94
96	-0.38
118	0.5

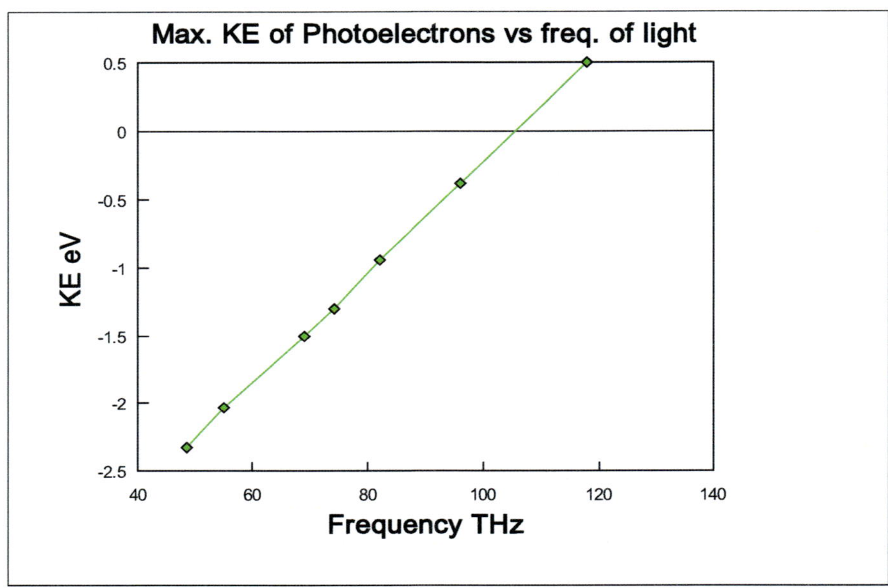

Figure 1.10 Maximum KE of photoelectrons vs. frequency of incident light. The maximum kinetic energy is proportional to the frequency and not related to the intensity of light incident because high frequency light waves affect the electrons of the inner shell orbits. [Figures from R. A. Milikan, Phys. Rev. 7,355 (1916) reproduced with permission of the University of Chicago Press.]

Further, it was claimed that the time-delay between initial light illumination and the onset of photoelectron ejection current, is so short **3×10^{-9} s(< 10^{-8} sec)** that it cannot happen with the wave model of the light. The onset time delay is still very high, compared to the period of light incident wave, which is a few femto seconds **(1.25-2.5 Fs**, see Table 2.1 for the visible light range). It is highly probable that photoelectrons may be ejected after it absorbs wave energy over several cycles of light waves. However, Einstein and the proponents of the particle theory of light, claim that photoelectrons are ejected by the energy absorption from exactly one quantum of photon. Even if that is a correct observation, it is not right to specify that the photon is a particle. Our conviction is that a particle should have rest mass and center of gravity [37]. Hence, the short delay does not prove that light consists of photon particles. In section 1.5.2, we will discuss the details of the short wavelength limit X-ray experiment. Again we will show that the results of that experiment do not prove that light consists of particles.

1.5.2 Short Wavelength Limit X-rays

The focus of this section is to describe the details of the short wavelength limit X-ray experiment. In this experiment, the metal target is struck with the high energy accelerated electrons, in the range of **5-50 KV**, a wide range of wavelength X-rays are emitted. The radiation is caused by the absorption of energy from the colliding electrons. The energy is absorbed by the electrons in the outer orbits and re-emitted in the form of X-rays. It is also possible that the X-rays are emitted because of the collision between the free electrons and the bombarded electrons. That is one reason that the sharp cut-off frequency only depends on accelerating potential and does not depend on the type of metal target. The work function does not appear because it is almost negligible in comparison to the electron and photon energies involved (level of 0.1%). Figure 1.11 illustrates the radiation event. Figure 1.12 plots the relative intensity of the X-ray emission vs. the wavelength for varying electron potentials; the corresponding numeric data is included in Table 1.2. Figure 1.13 illustrates the maximum frequencies vs. accelerating voltage; the corresponding numeric data is summarized in Table 1.3.

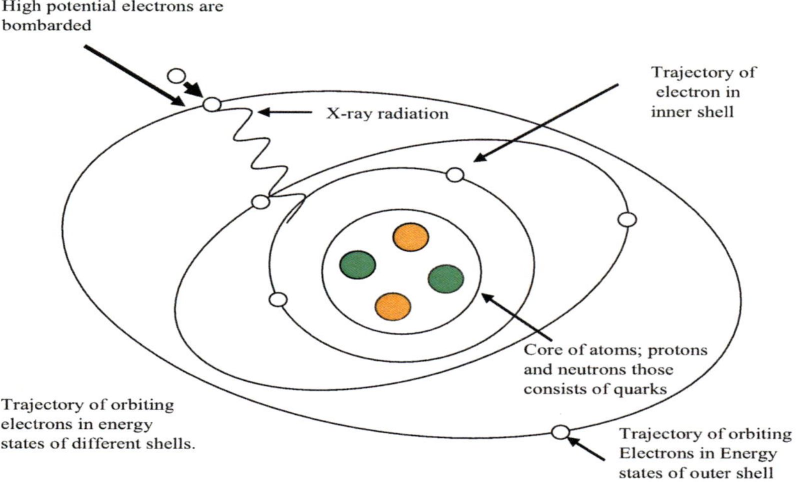

Figure 1.11 Radiation event caused by collision of high energy electrons. Here, it is demonstrated that an electron with high accelerating potential may trigger an orbital transition event and release X-rays.

The X-ray spectrum indicates the sharp-cut off at a minimum wavelength (or maximum frequency), that is the same for all metals and varies linearly with accelerating potential [3]. The phenomenon resembles the photoelectric effect in reverse. It is claimed that the maximum possible frequency of X-rays, corresponds to the kinetic energy of an incident electron, which is converted into the energy of a single photon. There is an error here. The highest potential electron produces the highest frequency X-rays, because the electron penetrates into the deeper shells of the electron orbits. The electrons collided in the orbits closer to the nucleus, have a higher speed and spin momentum, than the electrons in the outer shell orbits. As indicated in Figure 1.11, the excited electron absorbs energy from the colliding electrons and re-emits the excess energy in the form of X-rays. The maximum frequency of an X-ray, is the function of the excess energy and the quantum state (spin momentum and speed) of the electron from which the energy is released. The fact that this maximum frequency is linearly increasing with the accelerating potential of the electron, does not prove that the released X-ray is a photon particle. The energy content of a wave can be a quantized number just as well as that of a particle. Further, the spectra of the various wavelengths are emitted in this event. Therefore, the conclusion that the X-ray photon is a particle in this event is not accurate.

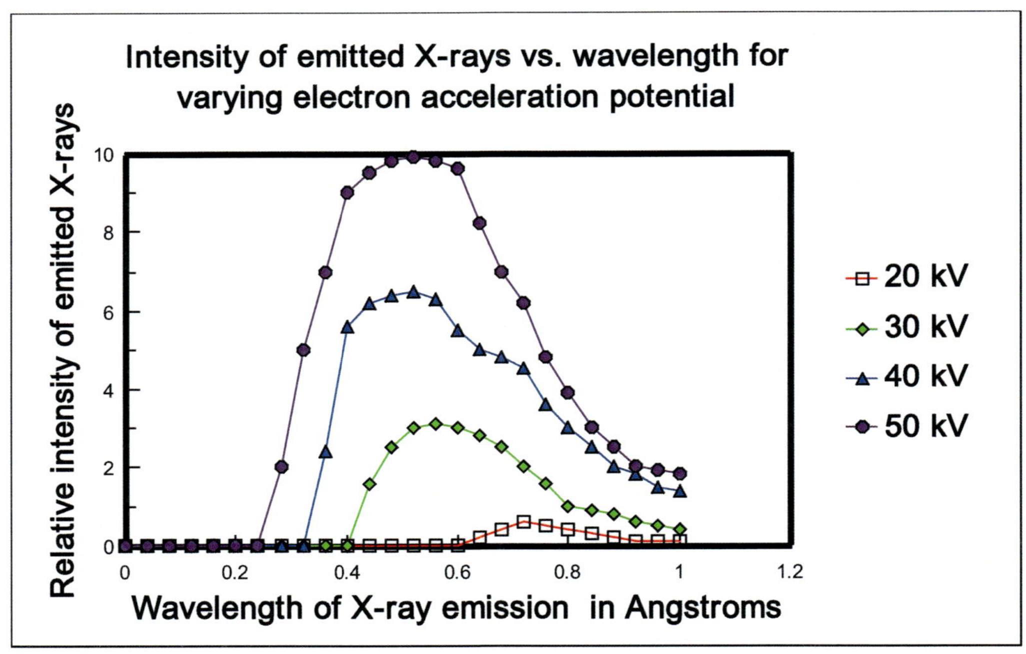

Figure 1.12 Bremsstrahlung spectra: electrons of various energies striking a metal target. X-rays are emitted when electrons with high acceleration potential are struck to a target metal. [Data of C. T. Ulrey, Phys. Rev. 11, 401 (1918),].

The phenomenon described in graphs resembles the Photoelectric effect in reverse, in the sense that kinetic energy of colliding electrons is transformed into X-ray radiation. It was discerned, that maximum possible frequency of X-ray corresponds to the kinetic energy of an incident electron which is converted into the energy of a single photon of the frequency. The energy from bombarding electron was absorbed by electron in orbit and re-emitted in the form of X-rays. Therefore, for a given accelerating potential, intensity of X-rays peaks at some wave length and is cut-off at some highest frequency. This frequency, at which peak intensity occurs, increases with increase in accelerating potential and so does the cut-off frequency. Further, it is established that the emitted X-rays waves do not possess rest mass and a unique position of center of gravity [37]. Therefore it is not correct to classify that radiated energy X-rays are comprised of particle photons of frequency corresponding to the X-ray energy waves. The distribution of rest mass in particle system defines instantaneous position of center of gravity for the mass in steady state. Thus wave entities such as light waves, infrared energy waves, X-rays, Cosmic rays and Laser or Maser waves should not be confused to be of particle type.

Table 1.2 X-ray emissions vs. cutoff frequencies at different eV acceleration.

Wavelength X-ray Ang.	Relative Intensity at 20 kV Acc.	Relative Intensity at 30 kV Acc.	Relative Intensity at 40 kV Acc.	Relative Intensity at 50 kV Acc.
0.0 – 0.24	0	0	0	0
0.28	0	0	0	2
0.32	0	0	0	5
0.36	0	0	2.4	7
0.4	0	0	5.6	9
0.44	0	1.6	6.2	9.5
0.48	0	2.5	6.4	9.8
0.52	0	3	6.5	9.9
0.56	0	3.1	6.3	9.8
0.6	0	3	5.5	9.6
0.64	0.2	2.8	5	8.2
0.68	0.4	2.5	4.8	7
0.72	0.6	2	4.5	6.2
0.76	0.5	1.6	3.6	4.8
0.8	0.4	1	3	3.9
0.84	0.3	0.9	2.5	3
0.88	0.2	0.8	2	2.5
0.92	0.1	0.6	1.8	2
0.96	0.1	0.5	1.5	1.9

In next section we will illustrate the details of experiment the Compton Effect, discovered and named for its inventor, Arthur Holly Compton. Compton Effect demonstrated that photon carries kinetic energy and a linear momentum. Collision event between X-ray photons and free electrons can be analyzed using the energy and momentum conservation laws of relativistic particle dynamics. In 1919, Einstein concluded that a photon of energy E travels in a single direction and carries a momentum = **E/c = hf/c.** In 1923, Arthur Compton (1892-1962) and Peter Debye (1884-1966) independently carried Einstein's idea of photon momentum further. Compton showed that when X-ray photon collides with free electrons; the X-ray photon suffers a loss of energy. The loss is manifested as an increase in wavelength of the X-ray by precisely the amount, corresponds to an elastic collision between two particles. The resulting scattering of X-rays and recoil of electron an effect resembling to collision between two particles is known as the Compton Effect. Even though many physicists thought that ideas developed from results of Compton Effect experiment were sufficient to prove particle nature of light, we think it is a wave.

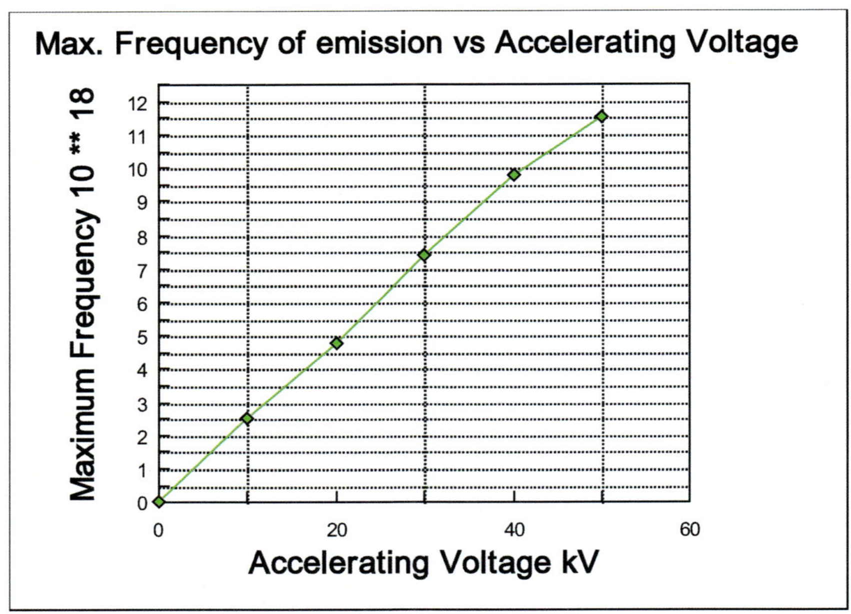

Figure 1.13 Maximum frequency of emission vs. accelerating voltage. The maximum frequency of emitted X-rays is directly proportional to the accelerating potential of bombarding electrons. [Data of C. T. Ulrey. Phus. Rev. 11, 401 (1918).]

Table 1.3 Maximum frequency of X-ray vs. the accelerating potential of electrons.

Accelerating Voltage V (kV)	Max. Frequency 10 ** 18 Hz
0	0
10	2.5
20	4.8
30	7.4
40	9.8
50	11.5

1.5.3 The Compton Effect

In this section, we will illustrate the details of the experiment that describes the Compton Effect. The Compton Effect demonstrates that a photon carries radiant energy and linear momentum. The collision event, between photons and free electrons, can be analyzed using the energy and momentum conservation laws of relativistic particle dynamics. Compton showed that when the X-ray photon collides with free electrons, the X-ray photon suffers a loss of energy. The loss is manifested as an increase in the wavelength of the X-ray by the precise amount that corresponds to an elastic collision between two particles. Here, Compton faced a dilemma. He stated that the X-ray was a particle. Then, he said that an elastic collision between the X-ray and the electron particle resulted in the increase of the wavelength of the X-ray. He implied that the particle X-ray was analogous to the wave X-ray, with a difference in wavelength caused by the conservation of momentum. This is interesting, but it does not prove that the X-ray is a photon particle. It is hypothetical that the X-ray photon collides with an electron. It is more appropriate and accurate, to specify that the X-ray radiation encountered a free electron. The electron absorbed the X-ray energy and re-emitted the energy at a lower wavelength. Again our proposition is that X-ray wave emitted carries energy but it has no rest mass [37]. Therefore, radiation waves X-rays do not have center of mass. Hence this experiment fails to prove that light or X-ray consists of photon particles.

Our explanation about the results of these experiments, clarifies that results do not necessarily prove that light energy consists of photon particles. In fact, the entire spectra of signals and radiation, uniformly consists of waves with different wavelengths and is completely characterized by the wave properties. In the following section, we will focus on the experiment, as well as the results, in order to show that light is indeed a form of wave energy.

1.6 Light is a Wave: Proof and Experiments

In section 1.5.3, we proved that the experiments were performed to validate that light exhibited particle behavior in a certain situation, is wrong. In this section, we will describe the facts and experiments to prove that light behaves as waves in all instances. We will validate our conclusion by citing several illustrations.

1.6.1 Doppler's Shift

The most remarkable distinction between waves and particles is that particles can't exhibit Doppler's shift. It is well known that a reflected signal, from a mobile target when recovered, has phase shift and or has shifted in frequency from the frequency of the original signal. This fact was discovered by an Austrian physicist Christian Johann Doppler in 1842. He observed that the sound waves emitted from a receding or approaching train has a different pitch when heard by a stationary observer. The pitch is higher when the train is approaching the observer and is lower when it is receding. A similar phase and frequency shift behavior is also detected for the entire electromagnetic wave signals. The Doppler's effect is consistently observed for signals of sound energy, light wave energy, X-rays, and infrared signals. The majority of scientists believed that Doppler's shift is the actual shift in the wavelength of the reflected signal. In reality, only an apparent shift in frequency is observed. The wavelength of the reflected signal does not change. It is well known, that the frequency shift is directly proportional to the velocity of the target. This is Doppler's popular principle that allows you to measure the velocity of a target object by measuring the frequency shift.

One of the common mistakes is that the wavelength of a traveling wave, changes when the source of the wave is moving, with respect to the medium of propagation. According to Doppler, the frequency or tone originating from the whistle of a moving train is sharper than when it is at a standstill. He measured the shift in frequency and determined that it is proportional to the speed of a train. This shift is popularly known as the Doppler shift. We believe that the Doppler's effect should not affect the wavelength, because the same shift in frequency should occur if the train is standing and the observer is moving toward the train. Therefore, the Doppler's Effect results in the change in the relative speed of the traveling wave front. The sound appears to be arriving at a higher speed and with a sharper tone, than if both the observer and the train were not moving. This is a very important observation, for light and other radiation energy waves, the fact that the Doppler shift affects the wavelength is a great source of error. The process for radiant energy source is the vibrations that are generated due to changes in the energy of the electrons when the transition occurs from one stable quantum state to the other quantum state. According to Niels Bohr's atomic model, these changes in energy are quantized for each different element and the shell of the electron to which it belongs. Therefore, the wavelength of light waves created, due to the disturbance of the state of fundamental particles, should not be affected by the macroscopic Newtonian motion of the molecules. Furthermore, the classical Doppler shift of a light ray, traveling at an angle $\boldsymbol{\alpha}$ to the direction of motion for two inertial systems moving with the speed $\mathbf{v}$ with respect to each other, is expressed by $\boldsymbol{\nu^* = \nu\,(1 - \cos\alpha(v/c))}$ [7].

We think that the Doppler shift is virtual shift. The reason is that a source of light from a star can't generate the different frequencies, if measured from two planets just because they are moving with different speed and in a different direction with respect to the star. For light waves, the Doppler shift is characterized as blue shift or red shift, depending on whether or not the source is approaching (increase in frequency) the observer or receding (decrease in frequency) from the observer [4]. Also, the fact remains that the measured Doppler shift has two values depending on the direction of the speed of the source in relation to the observer. A source of signal cannot create signals of two distinct frequencies at one instant which differ by $\mathbf{2\delta}$. According to Einstein's theory, the measured lengths of rods in inertial systems, is always smaller than the measured length of rod in a base reference system. As explained, the Doppler's effect predicts the high and low frequency values. Therefore, STR is not consistent with Doppler's effect. Scientists use these shifts to determine the distances of stars and the age of the universe. We will discuss the consequence of red shift/blue shift affecting the speed and not the wavelength of light, in Section 1.6.3.

Another example of the Doppler shift is its application to radar technology. Scientists measure the shift, in the frequency of the reflected signals received from a mobile target and the time of flight of **~50GHz** radio frequency signals, to determine the exact location of the target. The Doppler shift is directly proportional to the velocity of the moving target. Scientists typically assume that the wavelength of the returned signal remained the same (second order effect on the computation). In reality, the wavelength of the received signal is different because of Michael Faraday's laws of induction and not due to Doppler's effect. In Section 1.6.2, we will explain how the result of the experiment, which is about the rotation of the fan blades with a semiconductor photo sensitive coating, that when exposed to light energy, proves that light is comprised of waves.

1.6.2 Rotation of Fan Blades and Skin Effect

The supporters of the particle theory proposed yet another experiment. They prepared a fan with very thin (light) blades. Then the blades were coated with a thin layer of photo sensitive material. The fan was exposed to light rays. They discovered that the blades started rotating in a direction that will indicate that photon particles from the light transferred linear momentum into a torque. A construction of this fan and the apparatus is illustrated in Figure 1.14. The result does not prove that light consists of particles. The reason is simple, the size of the photons with zero rest mass, is very small compared to the mass and the momentum change associated with the blades. Further, the fan blades would not rotate if not coated with photo sensitive paste. The designers of the experiments erroneously concluded that photons were colliding with the blades and transferred their momentum. Therefore, a closer examination is needed to analyze the rotation.

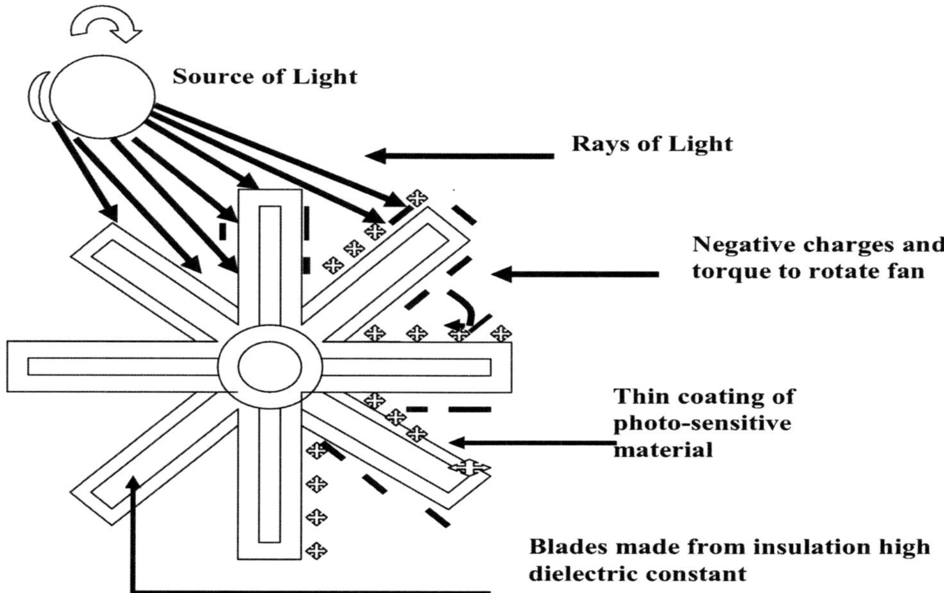

Figure 1.14 Rotation of fan blades coated with photo-sensitive material. When blades of a fan are exposed to light, charges on the surface of the fan blades are asymmetrically accumulated because of the skin effect and the absorption of wave energy. The fan sees a torque and rotates.

Figure 1.15 illustrates a detailed view of a pair of blades. The blades are made out of a very thin film of ceramic which has very high dielectric constant compared to air. Also, the thin film decreases the weight of the film and the whole fan, to less than an ounce. These blades are coated with photo-sensitive material. When the blades are exposed to an intense light beam, many electrons in the outer shell of the photo sensitive material absorb the light wave energy. They jump to a conduction band, become free electrons, and leave behind holes in the bound state. The free electrons accumulate on the surface of the blade because of the skin effect.

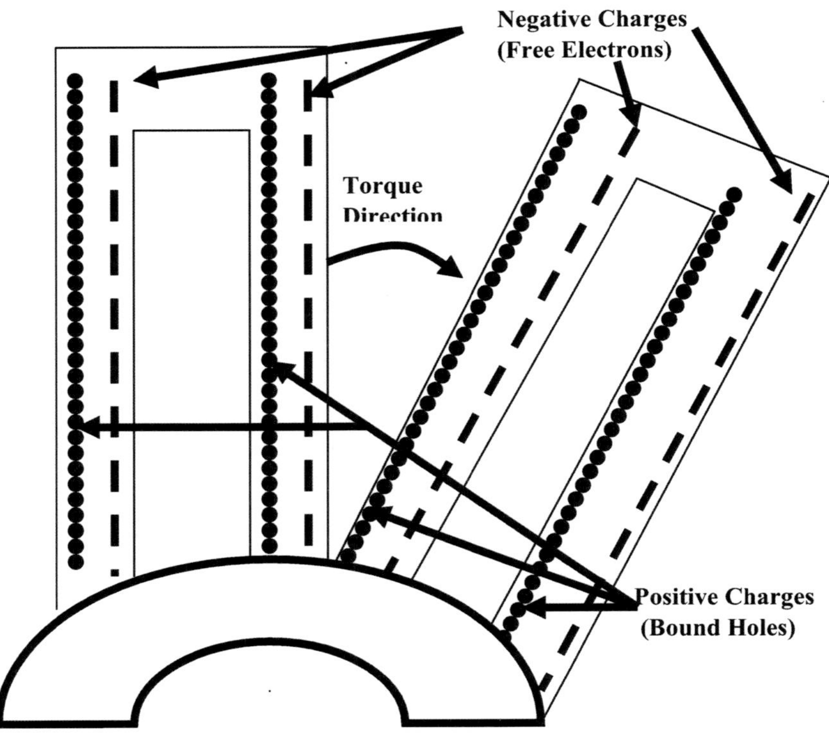

Created by Shailesh Kadakia on August 12, 2007

Figure 1.15 Details of the charges on a pair of fan blades exposed to a beam of light. A torque is developed when the thickness of the film that is pasted, is larger than the thickness of the fan blades.

The accumulation of negative charge, results in a potential difference and a static electric field. While charges attempt to balance charges of the opposite polarity, the force of attraction between the charges on two successive blades, creates momentum to realize torque. This torque is responsible for the rotation of blades. In reality, when the light waves hit the blades with a photo sensitive material coating, many (valance band) electrons are ejected because they absorb the energy from the waves. The freed electrons are in the conduction band and charge the surface of the blade negatively. The other side of the blades is positively charged because holes are created. The negative charge on one blade is attracted to the positive charge on the blade in front. The skin effect further increases the charge concentration on the surface of the blade, because blades are made out of thin film and the coating is very thin. The force of attraction between the charges results in the rotation of the blades and the fan. Thus, this experiment does not prove that light is comprised of photon particles. Instead, it proves that it is a wave. In sections 1.6.3 and 1.6.4, we will show more examples that substantiate our conclusion that light always behaves as energy waves, and never as particles.

1.6.3 Bright and Dark Sides of the Planet

We all know that the side of the Earth facing the sun is illuminated and day time for that region of the Earth. On the other side of the sun, it is dark and night time for that part of the Earth. If light is comprised of photon particles, as described by Einstein, it should be possible to bend the rays of light under the influence of some kind of force field, to illuminate the dark side. It may be a very tiny fraction, such as a few feet. This is an extremely difficult experiment to perform.

A major deficiency of the particle model for light stems from the fact that white light from the sun consists of the seven primary colors of the rainbow. There is no explanation about how the photons correspond to each color, with different frequency and wavelength, can be combined into photon of white light. Also, it is difficult to explain, based on the particle theory, how the photons of white light, when passed through a prism, can be split into photons of different colors. Splitting white light into different colors is very easy to observe naturally. For instance, daily, at sunrise (dawn) and sunset (evening), we see the red colored light more distinctly than the other colors. The index of the refraction is different for the red color. The interesting fact is that in the morning, that section of the Earth is approaching the sun. In the evening, that section of the Earth is receding from the sun because of the Earth's axial rotation. An observer should see the blue shift in the morning and the red shift in the evening. The shift is observed and verified by Michelson's apparatus experiment. Therefore, this phenomena of red colors of light at sunrise and sunset, proves that light is a wave.

If light is a particle with zero mass, then why should it be visible as a different color at different frequencies or wavelengths? This proves that light is a wave. All of the chemical or electrical properties of event related light can be explained by strictly stating that the light is a wave similar to electromagnetic, X-ray, and infrared waves. The bending of light through the atmosphere and through high density material, such as fluids, like water or oil, or solids, like glass, the human body etc., can be more easily explained by theorizing it as being a wave. The same is true for the reflection of light from shiny surfaces. Another example is fiber optic cables. Light can travel through the curves and angular bends of the fiber optic cables. A wave can create a curved path by multiple internal reflection phenomena. If light was a particle, it would not appear at the end of the tunnel of the fiber optic cable. In Section 1.6.4, we will talk about more examples that provide proof that light is behaving as a wave and not particle.

1.6.4 More Examples

Several phenomena were investigated to explain the dual behavior of light being a wave or a particle. For instance, one of the controversies in the earlier years, that light is a particle, was from photo voltaic cells which were producing an electrical voltage difference, by exposing those cells to the sunlight. This is not true because the wave energy can dislodge the electrons from the photo sensitive material, to create the potential differences between the electrodes of the photocells.

The second example is the use of monochromatic light beams, in etching film with light sensitivity. The etching of thin films is very commonly utilized in the fabrication of semiconductor circuits. With the advent of ultra large scale integration (ULSI) technology, it is possible to analyze the nature of etched surfaces, using various techniques. The photographs taken by SEM, TEM, AES, and other means, distinctively show the side encroachment (under and over) in the etch processes. The surfaces etched, are very much analogue in nature, or continuous, from the defect point of view and are smooth and round shaped. Light were a particle, the edges would be rough like a microscopic corrugated surface.

We would like to discuss two more examples, to prove that light is a wave, not a particle. In popular shows of musical dancing laser beams intersect in the sky. If one carefully notices, the path of the beam after the intersection does not deviate. The beam continues to travel in a straight line. Also, the color characteristics of the beam remain the same. If the light or photons were particles, the trajectory of the intersecting beams should be affected. We are suggesting that an experiment be conducted in which one should analyze the wave length of the beam of light after the intersection. Our belief is that the wave length of the light beam will be the same as that emitted from the source, when measured after the point of intersection. To improve the contrast and facilitate the measurement, two different colored sources of the light beam should be used.

1.7 Summary

In this chapter, we began with a discussion regarding the nature of light, a wave or a particle. Then, we described Einstein's experiment of light trajectory in an accelerating elevator. We demonstrated that the path of light should not bend by the force of gravity from the Earth. Next, we illustrated that when light from the distant star arrives on Earth, after passing by the sun, it will travel in a perfect straight line. Our conclusion is based on the observation, that the light from the star may encounter deflection from the gravity force of multiple stars. If light trajectory will be affected by gravity from all the stars, in that case the correct position of the distant star will be impossible to determine in real time. Our universe consists of billions of stars which are constantly moving. The effective gravitational field is indeterminate at any time. Next, we described Michelson's interferometer experiment to measure the speed of the Earth's rotation. One purpose of the experiment was to prove that the speed of light is constant. We explained why his experiment is not adequate to prove that the speed of light is constant. Further, we discovered that Michelson's interferometer is not the correct experiment that allows you to measure the orbital speed of the Earth surrounding the sun. After that section, we disclosed the details of Sky's Interferometer. This is an improved design of Michelson's interferometer apparatus. Our experimental setup has an advantage over Michelson's apparatus. We are relying on the fact that we will measure the path delay in one way trip of light and not the round trip delays as suggested in the Michelson's original apparatus. The results of our experiment allow us to measure the speed of the Earth's rotation surrounding the sun.

In the next section, we described the analysis of the data from three experiments that were designed to prove that light behaves as particles in those instances. A careful analysis suggests that the results from those experiments did not provide conclusive evidence that light waves are particles. Then, we described the facts and the experiment of the rotation of fan blades coated with photo sensitive paste. The results from the experiment substantiate the fact that light behaves as a wave. In the next chapter we will look at the physical properties of waves that distinguish them from particles. Then, we discovered that the mass to energy conversion expression $\mathbf{E} = \mathbf{M} \times \mathbf{c}^2$ is not sufficiently accurate to calculate the energy release during a nuclear radiation event. We will derive more exact expressions for energy release in a nuclear radiation event, based on the release of binding energy, a predominant cause. Also, we will show a different method of specifying the energy content of light waves, instead of the conventional expression $\mathbf{E} = \mathbf{h} \times \boldsymbol{\nu}$, here $\mathbf{h}$ is Plank's constant and $\boldsymbol{\nu}$ is the frequency of light wave energy. We shall close the chapter by specifying a new technique to characterize the energy content of light waves.

From examples described in this chapter, and after careful analysis, our conclusion is that the photon is a particle only in the abstract and theoretical sense. In the physical sense it exhibits wave properties more dominantly than particle properties and it is safe to say that light is a wave and not particle. Light beams should not bend under the influence of a force field, such as gravity, because only the vector force waves, such as electromagnetic and electrostatic waves, have the effect of the corresponding force field. Therefore, the trajectory of light beams in an accelerating elevator, and described in Chapter 4, of collapsing clocks in the time dilatation experiment, should be modified. The fact that light does not bend by the force of gravity, is good news. It allows astronomers to accurately predict the position of stars, galaxies, and nebulae, with high accuracy. We provided strong evidence that light is a wave, by discussing the correct interpretation of many experiments that were performed to prove that light did behave as a particle under specific conditions. We substantiate our findings, by discussing the results of experiments performed by Newton, Michelson, Compton, and others. Our explanation provides strong evidence that light behaves as wave and not a particle. In Chapter 2 we will describes the properties that distinguish a particle from a wave. Also, we shall discuss the meaning of the true speed of light.

2.

Physics of Light and Electromagnetic Waves

In this chapter our focus is to explain the fact that the speed of light and the speed of the electromagnetic radiation are variable. Not only do the speed of light and the electromagnetic waves depend on the frame of reference, they also vary in different mediums, and the different types of energy waves have different speeds. The speed of the propagation of radio waves is different than **c**. We will show that the speed of propagation for energy waves in the entire range of frequencies of the electromagnetic spectrum, cannot be **c**.

In Section 2.2, we shall explain the true meaning of the constancy of the speed of waves in general and as it is applied to electromagnetic radiation, including light waves. In Section 2.3, we shall describe the concept of absolute time. In Section 2.4, we shall discuss the passage of light through Newton's prism and calcite crystals, to show that light energy waves, cannot be particles. Our conclusion is based on the results of the prism experiment and the path of polarized light through calcite crystals. In Section 2.5, we will describe the various properties that distinguish waves from particles.

According to present understanding, the energy content of the electromagnetic waves is computed by Max Planck's formula:

$$\mathbf{E = h \times \nu} \qquad \mathbf{(2.1)}$$

where **h** is Planck's constant.

The value of Plank's constant is $\mathbf{6.63 \times 10^{-34}}$ **joule-second**.

It is more commonly expressed as $\mathbf{4.14 \times 10^{-15}}$ **eV-s**. We believe that this formula is not valid for the entire range of frequencies of the electromagnetic radiation spectrum, a situation analogous to the speed of the propagation of electromagnetic waves. In particular, we will show that the formula should be different at very low frequencies of **1 Hz**, as well as lower than that value. Even the speed of electromagnetic waves at those frequencies cannot be as high as **c**. In Section 2.6, we will describe an alternate method of specifying the energy content of the electromagnetic waves.

In the next section, we shall concentrate on the computation of the energy release during the nuclear reaction. We will show that the formula for energy release computations, suggested by the most celebrated scientist, Einstein, $\mathbf{E = m \times c^2}$ estimates the value for the energy released during a nuclear radiation event in excess amount than in reality. The primary cause of energy release is the release of binding energy during a nuclear fission or a fusion reaction because atoms configuration is altered. Typically, when a radioactive element, such as an enriched ^{238}**U** nucleus, absorbs a slow moving neutron, a nuclear reaction occurs. The absorption of the neutron causes the instability in the atom that splits the atom into daughter atoms with vast quantity of binding energy release. An empirical expression for the binding energy release during a radio-activity is included in the section. Einstein's formula for energy release $\mathbf{E} = \mathbf{m} \times \mathbf{c}^2$ is good if the mass **m** is not the consumed mass, but is replaced by the change in kinetic mass. The kinetic mass is the mass equivalent of the kinetic energy. Also, the speed of light **c** should be a value that corresponds to the true meaning of the constant speed of light **c**. According to Bohr's model the light wave energy is released when the electron changes its quantum state. The difference in the energy levels of the electron quantum state corresponds to the light wave energy released. Therefore, **E** predicted by Einstein's formulae provides for the computation of only the light wave radiation energy. Next, we are interested in the understanding of the temperature variation inside the core of the sun. Scientists have predicted that the temperature at the core of stars, such as the sun is very high. The measured temperature is as high as $\mathbf{1.5 \times 10^7}$ **°K**. We believe that more data is needed to verify that such high temperature values indeed exist. We will show that there are practical limits for such high temperatures. Therefore, in Section 2.8, we shall discuss the nature of infrared radiation and the relationship of atomic vibrations, with the temperature of matter in active stars, such as the sun. In the next section, we shall describe the characteristics of the temperature profile at the core of the sun.

The main focus in this chapter is to prove that radiation energy is waves that have distinct properties and can't be confused as a particle. On the other hand, we also wish to prove that elementary particles such as electrons, protons, neutrons, quarks, collectively leptons and baryons can't be called waves. Though in some instances, their behavior may be better understood by the wave model instead of the particle model. For certain events, modeling the behavior of light as a particle, increase the ease of understanding the strange behavior of light.

2.1 The Speed of Light and Electromagnetic Radiation

Several years ago we had studied optics. We were taught that the velocity of light is constant **c** in vacuum and is related to frequency by the following equation:

$$\mathbf{c} = \mathbf{f} \times \boldsymbol{\lambda} \qquad \textbf{(2.2)}$$

where **f** is the frequency
$\boldsymbol{\lambda}$ is the wavelength of light
c is the speed of light

In general, the propagation speed of a wave is directly proportional to product of its wavelength and repetition frequency, and is inversely proportional to the propagation constant for the medium through which the wave is traveling. For light waves the constant is known as the refractive index of the medium. Its value is unity for vacuum. Our detailed analysis suggests that the equation (2.2) should be modified. Actually, the speed of the electromagnetic radiation will vary in different mediums. It will slow down by a factor of the refractive index of the medium. Essentially, the wavelength $\boldsymbol{\lambda}$ of propagating energy waves is scaled down by refractive index sigma (σ) of the medium [6]. The altered relation should be the speed of the electromagnetic waves inside the medium which should be expressed as:

$$\mathbf{V = (f \times \lambda) / \sigma} \qquad \mathbf{(2.3)}$$

where σ is the index of the refraction and the other parameters are the same as defined earlier

Notice that we have used symbol **V** instead of **c** for the speed of waves. We suspect that the speed of waves for the entire electromagnetic spectrum may not be constant and should not be equal to the speed of light **c**. In Table 2.1, the parameters, speed, wave-length, frequency, and refractive index for different light colors, are summarized.

Table 2.1 Light wavelength λ (10^{-9} m), frequency ν (10^{12}), period (10^{-15}) and coefficient σ_{water}

	WAVE-LENGTH λ		PERIOD	FREQUENCY ν	REFRACTIVE INDEX σ OF WATER	
COLOR	Angstroms	Nanometers	Femto sec.	Tera Hertz	Real Part	Orthogonal
Violet	3800-4300	380-430	1.266-1.429	790-700	1.345	2.11 e $^{-10}$
Blue	4300-5000	430-500	1.429-1.667	700-600	1.342	3.30 e $^{-10}$
Cyan	5000-5200	500-520	1.667-1.724	600-580	1.339	4.31 e $^{-10}$
Green	5200-5650	520-565	1.724-1.887	580-530	1.337	8.12 e $^{-10}$
Yellow	5650-5900	565-590	1.887-1.961	530-510	1.335	3.53 e $^{-09}$
Orange	5900-6250	590-625	1.961-2.083	510-480	1.333	1.41 e $^{-08}$
Red	6250-7400	625-740	2.083-2.469	480-405	1.331	3.48 e $^{-08}$

From common laboratory experiments, we know that a signal generator can produce a sinusoidal signal of frequency **1Hz** or a fraction of a hertz. These signals are generated by under damped RLC circuits with large time constant near one or more seconds. If one would transmit a signal of such low frequency over an antenna the signal should travel a distance of its wave length in one second. At frequencies below one hertz, the distance will exceed the distance corresponding to the speed of light. This shall contradict the premise that no signal can travel at a speed greater than the speed of light. We raised this question to Jaffer. He mentioned that at such low frequency, obviously the electromagnetic wave signal will not propagate and antenna may not radiate energy at that frequency with any noticeable amount of effective radiated power (ERP). We agree with his comments. This fact proves the general formula to compute the speed of electromagnetic waves:

$$\mathbf{c = (f \times \lambda) / \sigma} \qquad \mathbf{(2.4)}$$

These do not work well for very low frequency signals. In fact, we believe that the expression will be different for the signal frequencies below the carrier frequencies at which the radio frequency signal generators are designed.

Table 2.2 Full Electromagnetic Spectrum Table

Electromagnetic Wave	Energy E (eV)	Frequency F (Hz)	Wavelength λ (μm)
Long Electrical Oscillation	$E < 10^{-10}$	$F < 10^{4}$	$\lambda > 10^{10}$
Radio waves	$10^{-11} < E < 10^{-5}$	$10^{3} < F < 10^{9}$	$10^{11} > \lambda > 10^{6}$
Microwaves	$10^{-6} < E < 10^{-3}$	$10^{9} < F < 10^{12}$	$10^{6} > \lambda > 10^{3}$
Infrared rays	$10^{-3} < E < 2$	$10^{12} < F < 5x10^{14}$	$10^{3} > \lambda > 0.77$
Visible light	$2 < E < 3$	$4x10^{14} < F < 8x10^{14}$	$0.77 > \lambda > 0.39$
Ultraviolet rays	$3 < E < 10^{3}$	$7x10^{14} < F < 3x10^{17}$	$0.39 > \lambda > 0.01$
X-rays	$10^{2} < E < 10^{6}$	$10^{16} < F < 10^{21}$	$0.1 > \lambda > 10^{-7}$
Gamma rays	$10^{4} < E < 10^{8}$	$10^{18} < F < 10^{23}$	$10^{-4} > \lambda > 10^{-8}$
Cosmic rays	$E > 10^{8}$	$F > 10^{22}$	$\lambda < 10^{-7}$

In Appendix A, a table of bands of frequency, the allocation and corresponding wavelength of the entire electromagnetic wave spectrum from ITU is included. Clearly, it is evident from the rows of lower frequencies in the table that at very low frequency electric signals cannot travel at the speed of light. That is the main reason that FM and AM modulated signals from Radio stations are transmitted on a high frequency carrier. Therefore, we believe that electromagnetic waves below certain carrier frequencies can't travel at the speed of light. Thus, it is essential that the frequency allocation chart of the Electromagnetic energy wave spectrum, stated in Appendix A, should be partitioned into more than one chart. Therefore, we are including three charts. Table 2.2 displays parameters frequency, Wavelength and energy for full electromagnetic spectrum. In Table 2.3 a range of wavelengths for infrared, colors of visible light and ultraviolet rays is specified. Radio frequency divisions and band names are classified in Table 2.4.

The entire electromagnetic spectrum plays an important role in the semiconductor industry. Modern communications equipment that transmit and receive radio or microwaves, as well as encode and decode information, benefit from a multitude of IC's designed to do these things or support them in one way or another.

Table 2.3 Optical Radiation Spectrum Table

Optical Radiation	Wavelength λ (μm)
Extreme Infrared	$40 < \lambda < 1000$
Far Infrared	$6 < \lambda < 40$
Medium Infrared	$1.5 < \lambda < 6$
Near Infrared	$0.77 < \lambda < 1.5$
Red	$0.622 < \lambda < 0.770$
Orange	$0.597 < \lambda < 0.622$
Yellow	$0.577 < \lambda < 0.597$
Green	$0.492 < \lambda < 0.577$
Blue	$0.455 < \lambda < 0.492$
Violet	$0.390 < \lambda < 0.455$
Near Ultraviolet	$0.30 < \lambda < 0.39$
Far Ultraviolet	$0.20 < \lambda < 0.30$
Extreme Ultraviolet	$0.01 < \lambda < 0.20$

Light emitting diodes (LEDs) and laser ICs employ the emission of the optical radiation, as a result of the electronic excitation, to perform their intended functions. X-rays and gamma rays are vital to many analytical techniques and manufacturing steps used in the industry today.

Table 2.4 Radio Wave Spectrum Table

Band Name	Acronym	Frequency	Wavelength
Extremely Low Frequency	ELF	3-30 Hz	100,000 km - 10,000 km
Super Low Frequency	SLF	30–300 Hz	10,000 km – 1000 km
Ultra Low Frequency	ULF	300–3000 Hz	1000 km – 100 km
Very Low Frequency	VLF	3–30 kHz	100 km – 10 km
Low Frequency	LF	30–300 kHz	10 km – 1 km
Medium Frequency	MF	300–3000 kHz	1 km – 100 m
High Frequency	HF	3–30 MHz	100 m – 10 m
Very High Frequency	VHF	30–300 MHz	10 m – 1 m
Ultra High Frequency	UHF	300–3000 MHz	1 m – 100 mm
Super High Frequency	SHF	3–30 GHz	100 mm – 10 mm
Extremely High Frequency	EHF	30–300 GHz	10 mm – 1 mm

Further, we infer that the speed of light will be lower in a denser medium like glass than air. The speed of light, when it passes from air to glass or water and back to air, will be different inside the glass or water. Our observation is that after it passes and travels again in the air, the speed is the same as it was on the other side of the prism in air. Very simple optics experiment can be performed to prove that trajectory of light rays is readily changed by strong forces in existence between lattice structure of atoms of mediums glass, water and other fluids as opposed to the weakest force such as gravity. In Table 2.5, we have described the relative strength of the various forces on elementary particles that occur in nature in decreasing order of their strength [5].

From the table, it is evident that the deflection of light waves is several millions of times large as a result of the refraction which is caused by strong force of scattering of light in space between nucleons as compared to bending caused by weak force of gravity. This behavior of light can only be explained by its wave nature. Armand Fizeau and Léon Foucault measured the value of the speed of light **c** in water and air [7]. They confirmed that it is different (smaller) than its value in air. Both of them and Thomas Young were in favor of wave theory of light. Newton and Einstein favored corpuscular (particle) theory. Young revived wave theory by verifying the principles of interference to explain the colored rings and fringes in his double slit experiment. It is discovered very short wave length energy waves smaller than microwave and radio frequencies, radiations are completely insensitive to strong electric and magnetic fields. We suspect energy waves such as visible light and infrared radiations falls into this category may not be of electromagnetic nature. More experiments needed to be performed to verify this fact. This opens up a huge opportunity for modern day physicist to cease opportunity and explore further revealing mysteries of light waves.

Table 2.5 Particle Interaction Forces

Force Type	Relative Strength	Mediating particle	effect distance	comments Objects/range
Strong	1.0	Gluons	10^{-15} m or 1 fm	Quarks/Short
Electromagnetic	10^{-2}	Photons	10^{-13} or 100 fm	Coulomb $1/r^2$ /Long
Weak	10^{-5}	W & Z Bosons	10^{-18} m or 10^{-3} fm	Decay/Short
Gravitation	10^{-39}	Gravitons	Infinite	Planets & Stars /Long Galaxies & Nebulae

2.2 True Meaning for the Speed of Light c

Now there is an obvious question, what is the meaning of the statement that the speed of light or electromagnetic waves is constant **c**? The value is **2.997924.58 × 10^8 m/s** exactly. Our understanding states that the speed of light obeys the Galileo transformation for different moving inertial systems. In a preferred frame of reference where the distance between point of interest and observer does not change with respect to time, if the speed of light is measured, its value will be always the constant **c**. We infer that distance travel by light in this frame of reference in one second of time corresponds to the speed of light **c**. When we look at our universe, that frame of reference is not found because all of the celestial objects in the universe are constantly moving. The main cause is that the motion of objects is affected by gravitational force of the remaining objects in the universe. Luckily an extremely close approximation to such a frame reference may be found in the neighborhood of vast celestial objects. For instance, source of light and measuring instrument both located such that they are under influence of huge celestial object the Earth's gravity would experience uniform equal acceleration. Therefore, distance between these two objects may be considered invariant in finite time. To balance the force of gravity, celestial objects are revolving in orbits by their kinetic energy.

Next, we will describe how to measure the absolute speed of light **c** in a preferred frame of reference. In such a frame of reference, the speed of light is invariant with regards to space and time. Figure 2.1 depicts a situation in which the distance between the source of light and the observer is maintained constantly. To establish that the distance between Source **A** and observer **B** is constant, we shall maintain distance between another reference point **C** and source **A** and distance between **C** and **B** is constant and equal. Also, we insist that angle **ACB** is a right angle. Therefore, spacing **AC** = **BC** = $\mathbf{C/2^{1/2}}$. Then by Pythagoras theorem side **AB** spacing between source and observer **AB** is computed by $\mathbf{AB = (C^2/2 + C^2/2)^{1/2} = C}$. Now that we have defined the exact meaning to the true speed of light, it may be good idea to describe the speed of light as a phasor. A phasor is vector which has amplitude and a phase to describe the orientation. The phasor representation for a light wave is very convenient because it is a travelling wave in space and it is a time varying luminous field. The light wave characteristics are no different than the electrical charge field waves, a situation analogous to the electromagnetic waves radiated through an antenna. One of the difficult aspects of light waves is to control the phase of light waves at the source that generates light waves. Nevertheless, we shall represent a light wave speed phasor as a sinusoid. We shall state that amplitude of sinusoid **c** is the true speed of light and is constant in space and vacuum. Symbolically, the speed of light:

$$\mathbf{c} == \mathbf{C_t \, Sin(\omega t+\phi)} == \mathbf{C_x + j\, C_y} \qquad \textbf{(2.5)}$$

where $\mathbf{C_x}$ is a component in x-direction
and $\mathbf{C_y}$ is a component in y-direction

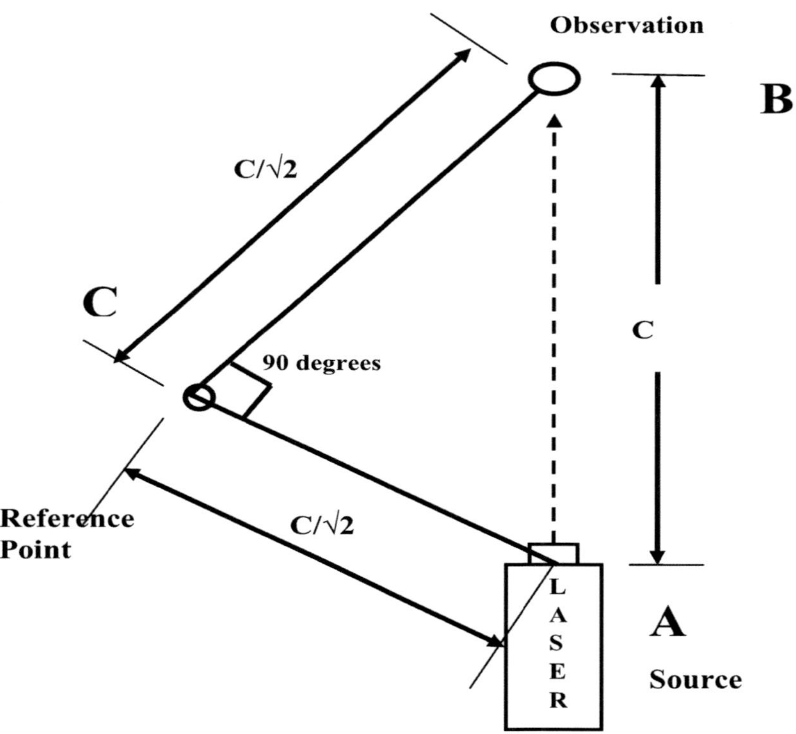

Figure 2.1 Measurement of True Speed of Light c. When the distance between the source and the observer of light does not alter with respect to time in the frame of reference, the speed of light measured is absolute

One of the most poorly understood concepts is the propagation of the force of gravity and how it interacts among the different objects in the universe at real time. It is not clear, at what speed the force of gravity travels in order to transfer its effect. It is believed that it is transferred by mediating particles graviton. There is no measured effect that verifies the existence of such particles. We shall discuss the speed of propagation of the force of gravity with more details in Chapter 7. It is worth to note that Einstein and other scientist work only refer to inertial systems moving with constant velocity with respect to each other. They have refrained from discussing the effects of the variation in the speed of light and electromagnetic waves for accelerating frame of reference. In reality as explained all celestial objects have either a circular orbit on a cross section of a sphere or an elliptical orbit a cross section of egg shaped surface, Circular or elliptical orbits means a state of constant acceleration. We believe further experiments are required to incorporate the effect of acceleration in moving frame of reference on the speed of light. It is evident, that the results of our measurement of the speed of light, heavily relies on the accuracy of time measurement. Therefore, in the next section, we shall explain the meaning of absolute time.

2.3 Absolute Time

In this section, we shall answer a very basic question what is an absolute measure for time. The concept of relative time measurement is obvious because one can easily determine difference in time between two events with very high accuracy. It turns out that absolute time definition is not simple and it cannot be determined independent of spatial distance. Therefore, to a first degree of accuracy one should find an event which is highly repetitive and the value of elapsed time for the event should not change for considerable length of time say several centuries. One such event is length of day on the Earth that corresponds to the period of the Earth's rotation surrounding its axis. It is determined that the length of an Earth day does not change by more than **1.0 ms** for the period of a year [36]. Since time on no atomic clock on the surface of the Earth can be independent of axial rotation and orbital rotation of the Earth the length of an Earth day should be considered as the most accurate time standard. Further, a second of time is precisely defined as **1/86400** of a day. Hence, the Earth's axial rotation period provides definition of absolute time by employing elapsed time measurement a unique event. According to recent statistics, in **4320000** sidereal years, the Earth rotates on its axis **1582237500** times and the moon rotates **57753336** times.

In the next section, we shall discuss passage of light through prisms and quartz crystals a experiment performed by Newton to prove that light consists of waves. We shall demonstrate that the wave model for light provides consistent results for all phenomena and obeys the second law of thermodynamics without any violation.

2.4 Passage of Light through a Prism

In this section, first we will describe the events of passage of light through prism. Next, we shall explain polarization of light energy waves. It is interesting to note that the kind of polarization exhibited by light waves is not observed for any other energy waves. Even though many scientists have classified light waves as electromagnetic waves, similar to electrical and magnetic energy waves, there is a distinct difference. Electromagnetic waves such as electric and magnetic waves are sensitive to poles. For instance, the field lines of magnetic energy starts from the North Pole and tends to end at the South Pole, and vice versa. The field lines of electric field static or time varying fields starts from a positive charge and ends at negative charge. This behavior of electrical and magnetic energy can be described as sensitivity to poles. Another way to put it is that the waves of this energy are polarized. The energy waves that show this field sensitivity tend to bend to complete their path. Therefore, it is possible that the electric energy or magnetic energy waves may bend by force of gravity. Although, there is no evidence that these energy waves indeed bend by the force of gravity. However, the light waves do not have sensitive poles so there is an even smaller chance that the electric or magnetic waves should be affected by force of gravity. We coined the term bipolarized energy waves, for the electric and magnetic waves, as each kind have bi-symmetric poles. The north and south poles for the magnetic fields, and positive **+q** proton charge or negative **-q** on the electron charge for the electric field. Since light waves have no pole sensitivity, but can have any plane of polarization, we shall call the light waves multi-polarized.

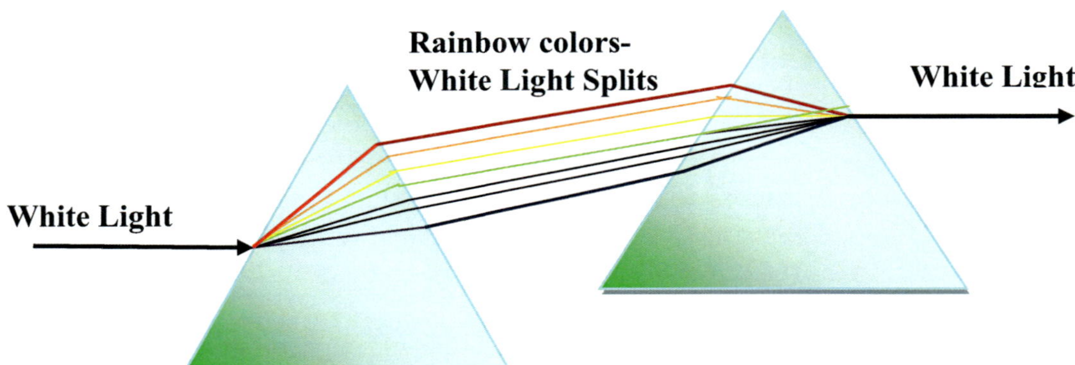

Figure 2.2 Passage of light through a prism pair. A beam of white light splits as it crosses the boundary of the medium from the air to the prism glass and recombines into white light when it crosses from the glass to the air outside the second prism.

Newton performed an experiment where he passed light waves through a pair of prisms. An arrangement for the set-up of his experiment is illustrated in Figure 2.2. The procedure and results of his experiments can be summarized as follows. He passed a beam of white light through a prism. The light coming out from the other side was split into seven colors similar to colors in a rainbow. The splitting occurs because the different colors of white light waves are refracted by the different amount in the glass prism. The propagation delay and the speed of the different colors of light, inside the prism is modified. The prism, which is made of glass, has a different index of refraction for each color corresponding to its wave length. The same event occurs when they exit to the air from the surface of the prism.

Next, when those seven colored light waves pass through a second prism they emerge as white light from other face of the prism. The splitting in seven colors and recombination in white light beam can't be accomplished perfectly every time if light is modeled as particle based on corpuscular (particle) theory. Further, the perfect recombination of the seven colors of light into one white light as it exits from surface of the second prism indicates that the entropy of the system of light particles and prism is decreased. This fact contradicts and violates the second law of thermodynamics that states, the entropy of any thermodynamic system always increases. The wave model of light does not violate the second law of thermodynamics because process of splitting white light into multiple colored light waves does not constitute entropy change. Therefore, white light must consist of the waves of colors, which are of different wavelengths and have different speeds in glass. However, the speed of waves of all colors must be the same in a vacuum. This experiment from Newton proves that the white light consists of waves of different colors.

While studying polarization effects of light, Etienne-Louis Malus discovered that polarization is not peculiarity of the light that have passed through a doubly refracting calcite crystal, but also may be produced by simple reflection [7, p.106]. Sir David Brewster showed that a light which has been reflected from a glass plate at a certain angle is reflected from a second plate by varying amount, if the latter is rotated about the incident ray. This implies that the ray behaves differently toward a second mirror according to position of the second plane of incidence to the ray from first. This behavior cannot be explained by the corpuscular theory, for a light particle striking the surface of the glass plate must either enter the plate or be reflected.

In recent times a property that plane of polarization of light can be changed when it passes through calcite crystals with certain lattice structures is utilized for data security [18]. The switching of plane of polarization for light waves corresponding to a bit pattern of ones and zeroes also known as quantum encryption is used as an encoding scheme to prevent hacking of computer data when it is transmitted over internet. It is interesting to know why quantum encryption provides a highly secured data link over the wired or wireless (Wi-Fi) internet connection. Because the information carried by the light energy waves cannot be decoded and retransmitted on a channel without adding any delays, it is impossible for a hacker to destroy the integrity of the transmitted data. Thus, the error-free data security is achieved when signals are transmitted and received over fiber optic channels. At present the length of such transmission is limited to a few meters. However, the length can be extended by inserting repeaters and amplifiers to restore signal strength against loss of attenuated signal power. The systematic response, or regularly exhibited phenomena of refraction, reflection, and the polarization of light through various types of material, proves that the light is a wave and not a particle. In the next section, we shall describe various properties of waves and particles that distinguish each other. It turns out that the most critical property of particle that distinguishes from waves is particles have a fixed center of gravity. This property many known physicist have failed to keep in perspective while discussing dual particle and wave nature and model of light.

2.5 Properties of Waves and Particles

Before one can determine if light a particle or a wave is, it is imperative that one should understand the differences between waves and particles. One of the basic differences is that particles occupy a finite volume in space, whereas waves propagate if not constrained. Further, to change the position, speed, acceleration, and other parameters of particles, an external agent such as force or torque is required. Also, the bound states of particles allow effect of external entity to change the position, velocity, acceleration and etc. parameters of the object with little difficulty. There is no way one could apply force or torque to alter position, velocity or other parameter of waves because they are propagating energy waves. Another remarkable distinction is that the parameters of a particle such as velocity, acceleration and position can be altered more readily than the corresponding parameters for waves. Also, the speed of light from a reflected surface is the same as the original speed **c** and the direction is reversed, without waves, the speed is diminished to zero. A particle must come to rest, at the surface when approaching, before its direction of movement can be reversed. This perfect property of elasticity can only be explained by wave theory. Further, a light beam can't be confined to some finite volume in space. After a light beam leaves its origin, one can't increase the intensity of the beam. One can't change the speed and other parameters of the energy waves. The single most important difference between particles and waves is particle have a fixed position for centre of gravity. Particles have mass that reflects quantity of matter contained in it. The position of center of gravity of particle with finite non-zero mass can't be changed without application of force. On the other hand, waves have energy but lack center of gravity. Waves propagate without application of force. Therefore, waves are distinctively different than particles.

At this point, we would like to discuss a perfect analogy. It is well known that other forms of energy waves exist in nature, such as electromagnetic, infra-red (radiation), sound (audio), and X-ray waves. All of these forms of energy are described more predominantly as waves. According to Joseph Weber and Friedrich Kohlrausch, the value for the speed of electromagnetic waves c is equal to that of the velocity of light **c**. James Clerk Maxwell concluded that the light waves are nothing more than electromagnetic waves. It is known that electromagnetic waves are analog in nature. Therefore, the particle behavior of light is inexplicable. The particle behavior is quantified as a discrete event. Practically all of the natural events are continuous. To digitize them, one uses the sampling process. This proves that the light waves are analog in nature just like electromagnetic waves. Another interesting fact about the light waves is that they do not have charge or magnetic pole strength, yet the waves consist of electric and magnetic field vectors that are normal in direction to each other. Further, the changing electric field lines produce varying magnetic field lines in a non-stationary, inertial frame of reference and vice versa. This means that the changing magnetic field lines in one frame of reference create a varying electric field in a non-stationary, other inertial frame. Why do light waves not exhibit any difference in the speed among the different frames? Also, is the speed of the electromagnetic waves of all frequencies the same constant speed **c** in different inertial frames of reference? The characteristic of light that changes in the speed of frame of reference, in which the light is propagating, does not alter any characteristic of the other frame under consideration. This property clearly distinguishes itself from magnetic and electric energy waves [38].

Let us suppose that visible light comprises of photon particles with velocity **c** and certain wave-length **λ**. The question is why laser and X-rays have such highly destructive effects. It is known that industrial lasers can cut through solid gold. Scientists argue that the laser is an energy beam with a very short wavelength. An astonishing fact is that the laser waves are also visible to the naked eye. If light is a particle, then laser beams must also consist of photons with a different frequency and energy level. The photons associated with the laser beams must possess a very high relativistic mass as compared to the relativistic mass of the photon associated with visible light. If we apply the postulates of the STR from Einstein, the effect of gravity on the laser beam should be several orders of magnitude higher than that on the light beam. Further, Einstein's theory does not specify that gravity itself changes the relativistic mass (sum total of rest mass and added mass due to kinetic energy) of photons or any other particles. Hence, the gravity should not cause the bending of laser waves.

From expression (1.4):

Relativistic mass for a photon associated with a laser of energy $\mathbf{E_l}$ **is** $\mathbf{M_l = E_l/c^2}$

Relativistic mass for a photon associated with the light of energy $\mathbf{E_v}$ **is** $\mathbf{M_v = E_v/c^2}$

A good application of laser technology is the development of weapons to destroy enemy targets utilizing laser beams. If the trajectory of laser beams were affected by the gravity of heavy terrestrial objects, such as the Earth, it would be deflected and require substantial correction. Otherwise, the laser beam would miss the target. In practice, no correction is required because laser beams propagate in a perfect straight line. Therefore, laser beams must be waves and not particles. Also, the main purpose of devising laser weapons is the fact that they can attack targets with great precision in a straight line.

This experiment can be performed in the modern world to verify the difference between the energy of the photons of light and laser. We believe that the laser and light energy behavior is identical in many respects as other wave energy forms such as electromagnetic waves at radio-frequencies (UHF and VHF). Intuitively, the trajectory of a wave should be affected by a force field such as gravity if the force is transferred through the medium to the wave in which the wave is propagating. However, the light waves do not require a medium to propagate. Furthermore, the force of gravity is defined between two objects with concentrated mass. Light waves have zero rest mass and its center of gravity is indeterminate for the kinetic energy mass. This statement is supported by the Heisenberg uncertainty principle which states that the product of uncertainty in position and the uncertainty in the momentum of a moving entity is greater than Planck's constant divided by **2π** [6]. Therefore, the path of a light beam should not be affected by gravity. Based on the analysis for light one can say that light behaves a lot more like a wave than a particle. Indeed light is a wave and its behavior should not be confused to be a particle. In the next section, we shall show a different method for expressing the energy content of light waves.

2.6 Energy Content of Light Waves

According to the physicist Max Planck, the energy of photon particles (light wave postulated as a particle by Einstein) is expressed as:

$$\mathbf{E = h \times \nu}$$

where **h** is Planck's constant in **J-s** and **ν** is frequency of light wave [4]

As per our explanation, light is a wave. Therefore, Planck's constant expresses the energy of one cycle of a wave at any frequency. This fact does not make sense. A photon with a higher frequency should possess higher energy than a photon at a lower frequency.

However, the energy of a light wave (photon), according to Einstein's STR is:

$$\mathbf{E = M_k \times c^2}$$

where M_k is the mass of wave due to kinetic energy

Equating both expressions for energy gives:

$$\mathbf{M_k = (\hbar \times \nu)/ c^2} \qquad \mathbf{(2.6)}$$

This indicates that the light waves have energy and corresponding to that relativistic mass. However, the rest mass for light waves is zero. The energy of a light wave should affect the trajectory of accelerated particles similar to the effects of magnetic field. We recommend that the effects of high intensity light beam on charged particle in a large Hadrons collider should be observed. The deflection of a charged particle with a light beam could possibly provide a very accurate means to measure the energy contents of light waves. The deflection should be directly proportional to intensity of the light wave which is expressed in units of lux and lumens in the MKS system and candle power in the FPS system. These measurements should provide another method of computing energy of a light wave. The energy content of electromagnetic waves is measured in terms of root mean square, average and peak values because voltages and currents causing the changes in time varying fields are sinusoid or combination of sinusoids in nature. An analogous procedure should exist for light wave energy because light waves are sinusoids.

Since the speed of light **c** is no longer constant and independent of a frame of reference, we shall revise the computation of the energy release from the nuclear reaction. As explained, the energy released during the nuclear reaction can be more accurately modeled as binding energy release. In the following section, we shall discuss an empirical expression for binding energy release.

2.7 Revised Energy Computations for Nuclear Reaction

According to Einstein's theory, energy released from radioactive material during nuclear reaction is computed from relation $\mathbf{E} = \mathbf{m} \times \mathbf{c}^2$ where **m** is consumed mass during nuclear reaction. This is not entirely true. The expression allows the calculation of a fraction of the energy converted into light waves. A large fraction of mass during nuclear reaction is converted into massive quantity of kinetic, thermal heat and sound energy. Also, very small fraction of consumed mass is transformed into light energy. This fraction is known as kinetic mass. As explained by Schrödinger [3], an electron in the high kinetic energy state is ejected to an orbit of low kinetic energy state when the light wave is radiated. As a result of this, the dynamic mass of the electron is decreased. Therefore, all material or rest mass fraction from the consumed mass is transformed into other non-light forms of energy.

As stated earlier, Einstein established that energy released during nuclear reaction is computed by relation:

$$\mathbf{E_n} = \mathbf{M_v} \times \mathbf{c}^2$$

where $\mathbf{M_v}$ is the consumed mass of radioactive material
and **c** is the speed of light in **km/s**

From our point of view this represents the maximum light energy that can be generated during the progress of the reaction. In reality, it is obvious that, along with light energy, massive quantity of thermal energy is generated. Also, tremendous amount of kinetic energy is released. Therefore, the energy computation for this reaction needs to be revised:

let $\mathbf{M_v}$ = total consumed mass in **kg**
$\mathbf{M_c}$ = mass transformed into light energy in **kg**
$\mathbf{M_i}$ = mass at the starting of the reaction in **kg**
$\mathbf{M_f}$ = mass at the completion of the reaction in **kg**

Then, mass transformed into energy forms other than light, can be found from the following equation:

$$\mathbf{M_v - M_c = M_i - M_f}$$

total energy = intrinsic energy + kinetic energy + potential energy + thermal energy

$$\textbf{total energy} = \mathbf{M_i \times c^2 + \tfrac{1}{2} M_i \times V_i^2 + M_i \times G \times D + M_i \times S_i \times (T_i - T_r)} \qquad \textbf{(2.7)}$$

The third and fourth terms are zero at the reference temperature and the referenced system.

Also, the energy equation is:

upper limit to energy release = actual light energy conversion + thermal energy transformation + increase in kinetic energy in the system

$$\mathbf{M_v \times C^2 = M_c \times c^2 + M_f \times S_f \times (T_f - T_i) + (1/2)\, M_f \times (V_f^2 - V_i^2)} \qquad \textbf{(2.8)}$$

where:
- $\mathbf{S_f}$ is the specific heat of the end product of reaction
- $\mathbf{V_f}$ is the final velocity in m/s
- $\mathbf{V_i}$ is the initial velocity in m/s
- $\mathbf{T_i}$ is the initial temperature in degrees K
- $\mathbf{T_f}$ is the final temperature in degrees K

Note that the thermal and kinetic energy forms are more catastrophic and causes more destructive effects than light energy. In fact, the light waves are harmless for the most part. Therefore, the emphasis should be on the energy generated in the forms other than light such as heat and explosion. However, majority of physicists emphasize the term on the left hand side. This is a highly exaggerated value and represents total energy content of matter before reaction expressed as light wave energy an expression proposed by Einstein.

It is found that after the completion of a nuclear reaction, the total number of protons, neutrons and electrons of the end products corresponds (same) to the count of constituent elements before the reaction. It is not clear how the mass is consumed and transformed into light wave energy except for the decrease in the mass due to the reduction in the kinetic energy of particles involved in the nuclear reaction. It is confirmed that during the course of nuclear reaction atoms are split at neutron, proton and electron level. The reaction does not cause smashing of protons and neutrons, which should result in release of quarks. Einstein's postulates described that consumed mass during nuclear reaction corresponds to energy $\mathbf{M_v \times c^2}$. However, this energy corresponds to decrease in kinetic energy of elementary particles. In reality, mass does not disappear during the course of nuclear reaction, except for the mass of the kinetic energy of the electrons, in the participating atoms. The mass of matter should be considered consumed if protons and neutron count of the atoms of initial element and final product element as a result of reaction differ. In our opinion, energy released by a nuclear detonating device is over estimated by Einstein's equation. Perhaps in the early 20th century, Einstein did not realize that matter consisted of the most fundamental particles, quarks or possibly strings, rather than electrons, protons, and neutrons, as described by modern physics. He assumed that protons, neutrons, and electrons were the basic particles. He also thought that those particles were changed to light and other forms of destructive energy when the mass was consumed during nuclear reaction.

The fact of the matter is that the light energy was released with the massive thermal energy, but the protons, neutrons and electrons remain intact during the nuclear reaction. In fact, the cause of massive energy created during a nuclear reaction is due to enormous quantity of binding energy from atoms of constituent elements is released. The binding energy release occurs because atoms configuration is changed in the course of nuclear reaction. As explained earlier, when a radioactive element such as enriched ^{235}U nucleus absorbs a slow moving neutron that is intentionally introduced by isotopes of ^{235}U, nuclear reaction occurs. The absorption of neutron causes instability in atoms nucleus. The nucleus splits into element atoms of smaller mass than before. This is an atomic fission process. In addition to splitting, vast quantity of binding energy is release during the course of the reaction. The formula for binding energy release is:

$$\mathbf{E_b = (Z \times m_p + N \times m_n - M \times A) \times 931.494\ MeV/u} \qquad \mathbf{(2.9)}$$

where:
Z = the charge number of the element $^{A}{}_{Z}X$ or the number of proton charges
N = the neutron number, or number of neutrons, in the element
A = the mass number = **Z + N**

also, **1u** (atomic mass unit) = $\mathbf{1.660540 \times 10^{-27}\ kg}$ such that $\mathbf{^{12}C = 12u}$

$\mathbf{E_{Ru} = 931.494\ MeV}$

Proton mass $\mathbf{m_p = 1.007276\ u}$
Neutron mass $\mathbf{m_n = 1.008665\ u}$
Electron mass $\mathbf{m_e = 0.0005486\ u}$

A nuclear fission reactor provides massive amount of natural energy source because to form a heavy nucleus from light constituent elements huge amount of energy must be supplied. When a heavy nucleus splits in fission process huge energy is released. Note that the total energy of a bound system is smaller than the combined energy of separate nucleons, Therefore, the expression for $\mathbf{E_b}$ in equation (2.9) is correct.

Also, a semi-empirical expression for binding energy release during a radio-activity is included [22]. For fission reaction involving elements with mass number **A > 15**, the formula is:

$$\mathbf{Eb = 15.7 \times A - 17.8 \times A^{2/3} - 0.71 \times Z\,(Z-1)/A^{1/3} - 23.6 \times (N-Z)^2/A\ MeV} \qquad \mathbf{(2.10)}$$

Notice that quarks are never produced in a nuclear reaction. Therefore, Einstein's expression for energy release estimated the generated energy computations in higher quantity than actual value. It is interesting to know that it is not possible to smash protons and neutrons into quarks by any ordinary means or they are not created even during complex chemical and nuclear reactions. The quarks are produced in the large super collider when the collision occurs between the heavy particle, a proton or a neutron, with the high speed of a light weight particle, an electron. The collision process is also known as annihilation. During this annihilation many types of radiations are produced including light waves. Therefore, there is no free process known which exists in nature that transforms rest mass of matter neutrons and protons into energy. If we could perform such a conversion it would solve all our world's energy problems.

Why it is not possible to convert a proton or neutron into energy? Is an excellent question raised by many modern physicists? The answer is provided by the conservation of baryon number law [2]. The term baryon is coined for particles proton and neutron those are known to be manufactured but are not found in nature freely. Almost all of the mass of ordinary matter is comprised of the baryons except for the mass of the orbiting electrons. The baryon's conservation law states that total number of protons plus neutrons in the universe cannot change. One can only transform one type of baryon into another kind during any reaction and maintain the proton and neutron count as constant. Because protons and neutrons can't be destroyed in practice, so the entire mass of any matter is not available for conversion to energy. This fact invalidates that expression for energy released during nuclear reaction $\mathbf{E_n = M_v \times C^2}$. Further, whenever a quark with large mass annihilates into quark with small mass, the reaction does not produce light instead an invisible particle neutrino is released [5]. In any of these events stable light wave energy is not radiated. Majority mass of heavy quark is transformed into lepton particles which has small mass. On the other hand, there is no process that occurs in nature which produces either proton or neutron by combining quarks and absorbing light wave or other energy. Because the life time of quarks is a few micro seconds, it is not easy to develop a process or technique to create heavy particles from quarks. It is known that light energy can be converted into electrical energy by exposing the solar cells to light. We have never expressed electrical energy as mass. Therefore, expressing light energy as mass equivalent does not make sense.

Scientists have predicted very high temperature at the core of stars such as the sun. The estimated temperature at the center of the sun is as high as $\mathbf{1.5 \times 10^7\ K}$. We believe that more data is needed to verify that such high temperatures values indeed exist. We will show that there are practical limits for such high temperatures. Therefore, in Section 2.8, we will discuss the nature of infrared radiation and relationship of atomic vibrations with temperature of matter in active stars such as the sun. Further, we will examine the boundary conditions and limits of high temperatures in core of stars such as the sun.

2.8. Infrared Energy and Vibrations in Atoms

For several years, scientists were puzzled, regarding the life span of an active star such as our sun. An obvious question is how long the sun will supply energy to keep life on Earth intact. At present it is believed that the sun and the entire solar system including the Earth was formed 4.5B years ago. It is speculated that the universe and the solar system was emerged as a result of huge explosion a big-bang event. With the help of background microwave radiation, physicists are concluding that the average temperature of our universe is **4K**. Also, it is verified by measuring z-shift of radiation arriving from the core of the sun that the temperature at the core of the sun is $\mathbf{1.5 \times 10^7}$ **°K**. The fact is also verified independently from the computation, by applying law of equilibrium between thermal gas pressure and gravitational pressure inside the sun [25]. One obvious concern is when will the fuel in the sun is depleted? At present, the energy on the sun is produced by the reaction of the thermonuclear burning of hydrogen into helium gas. To sustain this reaction, it is suspected that extremely high temperature is needed. In this section, we like to address the issue of estimation and measurement techniques for the temperature at the surface and at the core of the star sun. In particular, we shall explore what are the limits to which temperature of gas atoms could rise and is it realistic that the temperature of the sun is $\mathbf{1.5 \times 10^7}$**°K** as claimed by the current best estimates.

It is known that molecular and atomic vibration is manifested as temperature of any substance. Also, it is verified that amplitude of vibration is proportional to temperature. The larger the amplitude, of vibrations, higher is the temperature of substance. We tried to search data of amplitude of atomic vibrations vs. temperature of gases, liquid and solid substances. We did not find sufficient data for any fluids or gases. The most probable reason is that direct measurement of atomic vibrations is not straight forward process. Some indirect techniques need to be invented. Specifically, measurement for a broad range of temperatures from temperature of liquid nitrogen to high temperature values at the surface of the sun could be expensive and may require a lot of resources. Further, there is no direct procedure is known to date that allows measuring surface temperature of the sun. The temperature computations projected by thermal and gravitation equilibrium may not be correct. The reason is the density and pressure of the sun at the core will not permit the vibrations of atoms to realize high temperature of **15M°K**. We believe that the temperature rise caused by the thermonuclear burning of hydrogen into helium may be limited by inter atomic distance of helium atoms. Therefore, in the next section, we shall investigate the temperature profile of the sun based on present state of the art methods for temperature determination.

2.9 Temperature Profile of the Sun

Before we explore the details of process to compute temperature inside the core of the sun let us understand importance of prevailing high temperature need inside. We all know that existence of life and survival of our solar system heavily relies on the fact that the sun continue to provide radiation energy light and heat waves. As we shall show that our computations in Section 8.5 reveals that at the present rate of energy emission the sun loses mass of $\mathbf{6 \times 10^{11}}$ **kg** per second from the thermonuclear burning of hydrogen into helium [17]. Naturally, this loss of mass and loss of energy should affect temperature of the sun. At least some section of the sun will experience decrease of temperature. Also, one needs to know at what minimum temperature the thermonuclear burning of hydrogen into helium will not occur. Because those sections of the sun have a temperature below that limit, their hydrogen will not burn into helium to produce the high energy output. To predict the life span of a main sequence star such as the sun it is imperative to estimate temperature variation inside the sun and relative concentration of hydrogen gas from the sun's surface to the center. In the following, we shall first describe a method that provide very rough estimate of temperature at the center of the sun. Then, we shall discuss the technique that presents the temperature profile for the sun's core with high accuracy.

For a long time, the computation of temperature at the center of a star was based on the principle that to be in a steady state, the thermal gas pressure at the center should be in equilibrium with the gravitational pressure [28]. Without repeating the derivation, we shall utilize the final results to get a preliminary estimate on the temperature at the center:

$$\mathbf{T == G \times m_A \times M / (k \times R)} \qquad \mathbf{(2.11)}$$

where $\mathbf{T}$ is temperature in °K
$\mathbf{G == 6.7 \times 10^{-11}\ N\ m^2/kg^2}$ the constant of gravitation:

$\mathbf{m_A == 1.7 \times 10^{-27}\ kg}$ mass of hydrogen
$\mathbf{M == 2.0 \times 10^{30}\ kg}$ mass of the sun
$\mathbf{R == 2 \times 10^{8}\ m}$ radius of the sun
$\mathbf{k == 1.4 \times 10^{-23}\ J/K}$ Boltzmann constant

the substitution of parameters in equation (2.11) gives:

$$\mathbf{T == 2.3 \times 10^{7}{}^{\circ}K == 23M^{\circ}K}$$

the pressure in the center of the sun is:

$$\mathbf{P == \rho \times G \times M/R}$$
$$\mathbf{== 2.7 \times 10^{14}\ N/m^2}$$
$$\mathbf{== 2.7 \times 10^{9}\ bar}$$

with the mean density:

$\rho == \mathbf{1.4 \times 10^3\ kg/m^2}$

a more accurate result for **T** is obtained by using values that are close to the actual values:

$\mathbf{G} == \mathbf{6.67 \times 10^{-11}\ N\ m^2/kg^2}$ constant of gravitation,,
$\mathbf{m_A} == \mathbf{1.6735 \times 10^{-27}\ kg}$ mass of hydrogen
$\mathbf{M} == \mathbf{1.9891 \times 10^{30}\ kg}$ mass of the sun
$\mathbf{R} == \mathbf{6.96 \times 10^{8} m}$ radius of the sun
$\mathbf{k} == \mathbf{1.3806 \times 10^{-23}\ J/K}$ Boltzmann constant

the substitution of parameters in equation (2.11) gives:

$\mathbf{T == 2.310633 \times 10^{7}{}^{\circ}K}$

a more realistic value of the temperature at the center of the sun is:
$\mathbf{T == 1.55 \times 10^{7}{}^{\circ}K}$ or lower than this value is far more accurate

Table 2.6 A theoretical model for the Sun.

Distance from Center % Radius	Brightness Fraction	Mass Fraction	Temp. $\times 10^6$ K	Density Kg/m^3	Pressure Rel. to Cen.	Mean free path mm
0.0	0.0	0.0	15.5	160,000	1.00	0.243
0.1	0.42	0.07	13.0	90,000	0.46	0.443
9.2	0.94	0.35	9.5	40,000	0.15	0.992
0.3	1.0	0.64	6.7	13,000	0.04	2.624
0.4	1.0	0.85	4.8	4,000	0.007	10.742
0.5	1.0	0.94	3.4	3,000	0.001	53.261
0.6	1.0	0.98	2.2	1,000	0.0003	114.876
0.7	1.0	0.99	1.2	400	4 × 10-5	469.947
0.8	1.0	1.0	0.7	80	5 × 10-6	2193.086
0.9	1.0	1.0	0.3	20	3 × 10-7	15664.9
1.0	1.0	1.0	0.006	0.00003	4 × 10-13	23.498X10^6

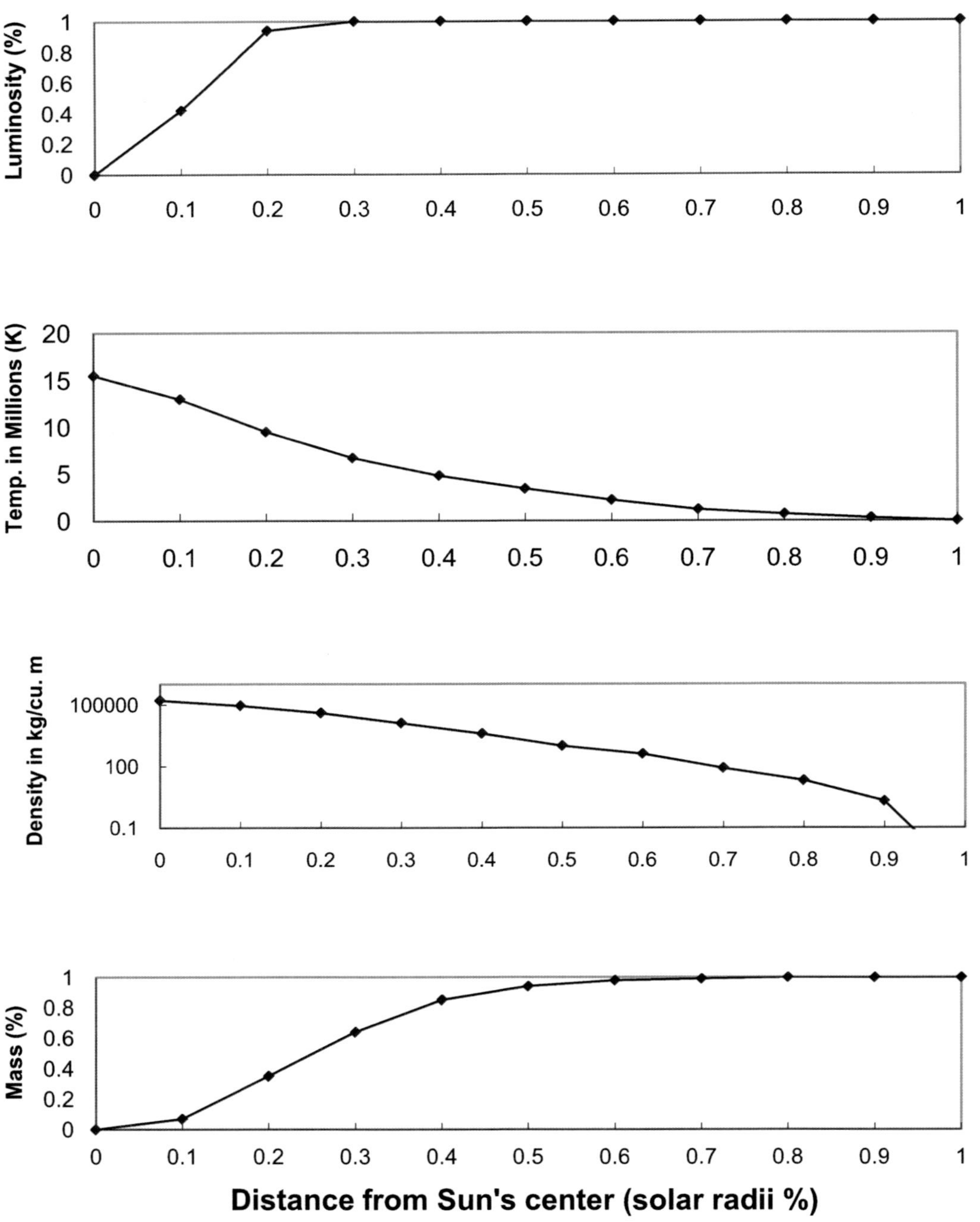

Figure 2.3 Theoretical model of the sun's interior. The parameters luminosity, temperature, density, and mass are plotted vs. the distance from the sun's center. (Courtesy of R. Freedman and W. Kaufmann III, Universe- Sixth Edition)

Now let us discuss the issues related to current theoretical model of the sun's interior. Specifically, we shall look at variation of parameters density, temperature, mass and pressure with respect to the distance from the center of the sun. In Figure 2.3, the models for the parameters pressure, mass, temperature, and density are reproduced from [17]. Table 2.6 depicts a numeric version of model data. From the data, it is evident that 85% of the mass of the sun is occupying about 40% of the radius volume of the sphere. In the same region temperature drops from **15.5M°K** at the center to about **4.8M°K** at the surface of sphere with a radius 40% of the full radius of the sun. We computed the mean free path (MFP) for hydrogen ions as a function of distance from the center using the formula:

$$\mathbf{MFP\ l = K_b \times T/(\sqrt{2} \times d_h^2 \times P)} \qquad \mathbf{(2.12)}$$

where **P** is pressure and $\mathbf{d_h}$ is diameter of hydrogen atom [31]

We believe that collision rate for mean free path above **10 mm (1 cm)** will not be high enough to sustain the thermonuclear fusion process. From the table, we find the minimum temperature at which the fusion process may occur is **5.0M°K**. We shall utilize this fact to estimate the fraction of available hydrogen mass for conversion into helium for the thermonuclear fusion process. This fractional mass will allow us to compute the life span of the solar system with higher accuracy than before.

Figure 2.4 illustrates the changes in the mass fraction of hydrogen and helium concentration at the interior of the sun, from the birth of the sun and at the present time. Figure 2.4a illustrates the hydrogen concentration at the time the sun was formed and its estimated value today. In Figure 2.4b, a similar chart for helium concentration is reproduced [17]. Because the He atoms are four times heavier than the hydrogen atoms, and the He concentration is 50% or higher in 10% of the radius from the center of the sun in a small region corresponding to 5% of the radius, it has a small or negligible thermonuclear reaction. Hence, the center of the core, comprised of a region less than 5% of the radius of the sun, may be cooler than **15.5M°K**, as predicted. Further, at a distance of 60% of the radius, the temperature drops to **2.2M°K** and the pressure drops to 0.03%. If we assume that the minimum temperature for the thermonuclear burning of hydrogen into helium is about **5M°K** only the mass fraction of the hydrogen mass from 5% radius to 75% radius should be considered as available for the thermonuclear burning. From the mass graph, this amounts to 70% of the mass of the sun. This should allow us to compute the remaining life span for the sun with high accuracy. We shall complete our discussion about the life span of the solar system in Section 8.5.

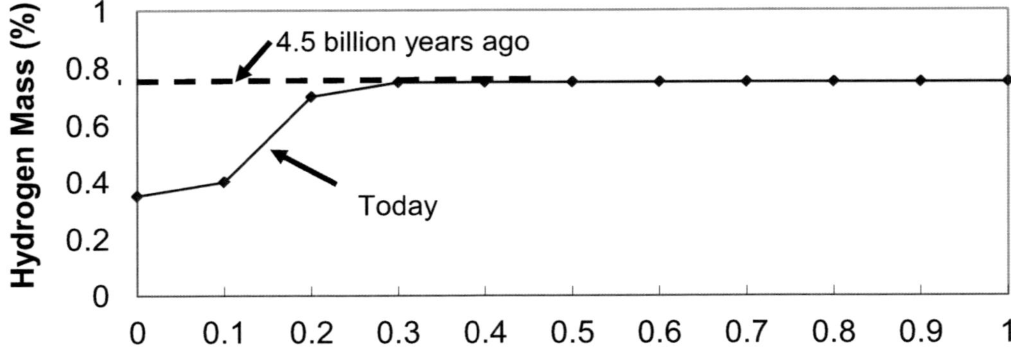

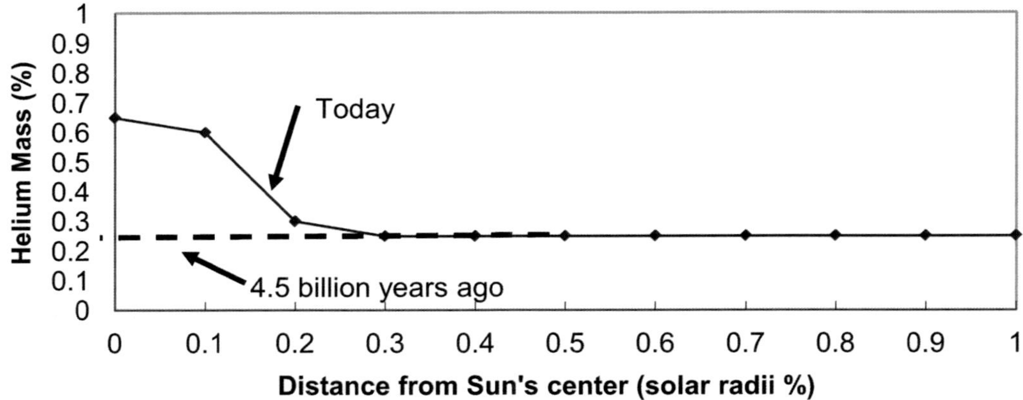

Figures 2.4a and 2.4b Changes in hydrogen and helium concentration within the sun. The proportion of hydrogen and helium concentration is changing because of the thermonuclear fusion of hydrogen into helium inside the core of the sun. (Courtesy of R. Freedman and W. Kaufmann III, Universe- Sixth Edition)

2.10 Summary

In this chapter, our focus was to explain that the speed of light and the electromagnetic radiation is variable. Not only do the speed of light and electromagnetic waves depend on the frame of reference, it also varies in different mediums, and the different types of energy waves have different speeds. The speed of propagation of the radio waves is different than **c**. In Section 2.1, we showed that the speed of propagation for the energy waves in the entire range of frequencies of the electromagnetic spectrum can't be **c**.

In Section 2.2, we explained the true meaning of the constancy of the speed of waves in general and as they are applied to the electromagnetic radiation including light waves. In the next section, we explain the meaning of the term absolute time. We concluded that absolute time can only be derived from a time invariant event such as the Earth's rotation surrounding its axis. In Section 2.4, we discussed passage of light through Newton's prism and calcite crystals to show that light energy wave cannot be particles. Our conclusion was based on results of the prism experiment and path of polarized light through calcite crystals. In the Section 2.5, we described various properties those distinguish waves from particles.

According to present understanding, energy content of electromagnetic waves is computed by Planck's formula $\mathbf{E = h * \nu}$ where **h** is Planck's constant. The value of Plank's constant is $\mathbf{6.63 \times 10^{-34}}$ **joule-second**. It is more commonly expressed as $\mathbf{4.14 \times 10^{-15}}$ **eV**-s. We believe that this formula is not valid for the entire range of frequencies of the electromagnetic radiation spectrum, a situation analogous to the speed of the propagation of electromagnetic waves. In particular, we showed that the formula should be different at very low frequencies of **1Hz** and lower than that value. Even the speed of the electromagnetic waves at those frequencies cannot be as high as **c**. In Section 2.6, we described an alternate method of specifying energy content of electromagnetic waves. In the next section, we concentrated on formula for computation of energy release during nuclear reaction. We showed that formula of energy release computations suggested by the most celebrated scientist Einstein
$\mathbf{E = m \times c^2}$ estimates the results in excess value. In fact, the primary cause of energy release is the release of binding energy during a nuclear fission or a fusion reaction because atoms configuration is altered. Typically, when a radioactive element, such as the enriched $\mathbf{^{235}U}$ nucleus, absorbs a slow moving neutron, nuclear reaction occurs. The absorption of neutron causes instability in atom that splits into daughter atoms with vast quantity of binding energy release. An empirical expression for binding energy release during a radio-activity is included in the section. Einstein's formula for energy released during nuclear reaction is good if the mass m is not the consumed mass but it is replaced by change in the kinetic mass which is not the rest mass, but the mass equivalent of the kinetic energy. According to Bohr's model, the light wave energy is released when the electron changes its quantum state. The difference in the energy levels of the electron quantum state corresponds to the light wave energy released. Therefore, **E** predicted by Einstein's formulae, provides for the computation of the only the light wave and radiation energy.

Scientists have predicted very high temperature at the core of stars such as the sun. The measured temperature is as high as $\mathbf{1.5 \times 10^{7}{}^\circ K}$. We believe that more data is needed to verify that such high temperatures values indeed exist. In Section 2.8, we showed that there are practical limits for such high temperatures. Therefore, in Section 2.8, we discussed the nature of infrared radiation and relationship of atomic vibrations with temperature of matter in active stars such as the sun. Importance of estimation of temperature profile at the interior of the sun is evident from our discussion in Section 2.8. Therefore, in Section 2.9, we analyze the temperature profile at the core of the sun. One of the applications of studying temperature at the interior of the sun allows us to predict the life span of the solar system and our civilization. Our detailed analysis on the present state inside the sun indicates that only 70% of the mass of hydrogen contained in the sun is available for the thermonuclear fusion process. We shall use this mass fraction to finish the life span computations based on the assumption that the thermonuclear burning process can't occur below a temperature of about **5M°K**.

The main purpose of our focus in this chapter was to prove that radiation energy is waves that have distinct properties and can't be confused to be particle. Also, we proved that elementary particles, such as electrons, protons, neutrons, quarks, collectively leptons and baryons can't be called waves. Though in some instances, their behavior may be better understood by wave model instead of particle and modeled with ease to its complementary wave nature.

In the next chapter, we shall review the postulates of Einstein's theory of relativity. Then we shall formally describe the new postulates of the theory of relativity, hence forth labeled **Skylativity®** theory, relativity applied to the sky. Next, we shall discuss differences between Einstein's work and the new postulates developed by us. At first instinct it might appear that this work leads to old theory of classical mechanics. Detailed analysis will reveal that this work fills the holes that were left by Einstein in his STR. Specifically; we shall describe the implications on the speed and nature of light when the light source is experiencing acceleration so as to speak of an accelerating frame of reference. Einstein left gap in his theory by not discussing effect of acceleration on the speed of light when the source of light is residing in an accelerating frame.

3.

Postulates: Relativity Basics

In this chapter, we shall review the postulates of Einstein's special theory of relativity (STR) and general theory of relativity (GTR). Then, we shall look at the new postulates of relativity, the **Skylativity®** theory. Occasionally, we shall refer to Einstein's theory of relativity as E-theory for convenience, and to **Skylativity®** theory as K-theory.

In Section 3.1, we shall review the postulates of the special theory of relativity. In Section 3.2, we shall describe the concepts of the general theory of relativity. In Section 3.3, we shall formally describe the new postulates of **Skylativity®** theory, or K-theory, relativity applied to the sky and outer space. In Section 3.4, we shall discuss the postulates of general K-theory, and describe the differences between Einstein's work and the new postulates. It might appear that this new work leads to the old theory of classical mechanics, but a detailed analysis will reveal that this work fills the holes those were left behind by Einstein in his special and general theories of relativity. Specifically, we shall address the issues pertaining to the implications on the speed and nature of light, when the light source is experiencing acceleration, in an accelerating frame of reference.

Einstein took it for granted that the speed of light must be constant and independent of frame of reference, a concept he borrowed from Maxwell. Maxwell's equations assumed that the speed of light does not vary, even if a frame of reference is mobile. Therefore, in Section 3.5, we shall see the ways that Maxwell's equations are affected by the changes in the speed of light. In Section 3.6, we shall discuss the changes required in Einstein's field equation to account for the variable speed of light.

Einstein left a gap in his theory by not discussing the effect of acceleration on the speed of light, when the source of light is residing in an accelerating frame of reference. He favored the kinetic energy effect to the potential energy effect, on the mass of an object. He stated that the total mass of an object consists of the rest mass plus the energy mass gained by additional kinetic energy mass, due to the speed of the object. As we shall see, mass of an object should be constant, a value that corresponds to the rest mass. The mass gained by the object, due to the speed changes in the object, is not a real mass change because the changes in the kinetic energy of the objects, is always compensated by the changes in the potential energy of the system of objects.

3.1 Principles of the Special Theory of Relativity from Einstein

In this section, we shall review the postulates of the special theory of relativity (STR) as stipulated by E-theory. Einstein formulated the principles of STR in 1905.

The principles of Einstein's special theory of relativity are based on the following postulates [7]:

- The laws of physics may be expressed in equations having the same form in all frames of reference moving at constant velocity with respect to one another. Not only the laws of mechanics, but those of all physical events, in particular, electromagnetic phenomena, are completely identical in an infinite number of systems of reference, which are moving with constant velocity relative to each other and which are called inertial systems. In any of these systems, lengths, and times, measured with the same physical rods and clocks, are different values in any other system, but the results of the measurements are connected with each other by the Lorentz transformation [19].

- The speed of light in free space has the same value for all observers, regardless of their state of motion. Einstein substantiated the constancy of the speed of light by the argument that the measured value of light $\mathbf{c} = (\mu_0 \times \varepsilon_0)^{-1/2}$ **m/s** for any frame of reference. Here μ_0 is permeability of free space and ε_0 is permittivity (a dielectric constant) of free space. He believed that μ_0 and ε_0 free space constants, should not change for a moving frame of reference. We shall explain later, in Section 3.3, why his assumption for permeability of free space, constant μ_0, is constant and independent of motion, which is not correct for a mobile frame of reference.

- Einstein stated that a quantity of mass, in principle, can be converted into an amount of energy $\mathbf{E} = \mathbf{m} \times \mathbf{c}^2$ where **m** is the consumed mass of material and **c** is the speed of light. Later in the 1920s, British astronomer Arthur Eddington and astronomer Robert Atkinson applied the principles of Einstein to calculate the energy released during the thermonuclear fusion of hydrogen atoms into helium atoms a process occurring within our sun. The fact that, in spite of the radiation of billions of watts of power per second from the sun, and resulting in a very small decrease in the mass, provided the first evidence that Einstein's energy transformation equation made sense. However, this translation of the mass into the energy equation, does not quantify the energy components into which the mass is transformed. It does not state how much of the energy in infrared, visible light or dissipated as an increase in kinetic energy of the system. Also, this energy equation does not state the phase relationship among the different components of energy waves.

- A changing magnetic field vector in one frame of reference, induces a changing electric field vector in a frame of reference that has relative motion, with respect to the former frame of reference and vice versa.

We disagree with the postulates of STR, in particular, as it relates to the Lorentz transformation. The Lorentz transformation assumes that the speed of light **c** is constant among different inertial systems, which is not a valid assumption. Also, we believe that the speed of light should be different for the observer who is traveling in space with the speed **v,** as compared to other inertial systems in which the speed of light is **c** from his frame of reference. The speed of light should be related by the Galileo transformation [7], **c' = c - v** for approaching, and **c' = c + v** [7, (42) 125] for receding. The situation is analogous to an aircraft traveling at a supersonic speed (the speed above the sound waves). In fact, a large sonic boom occurs when the nose of the aircraft penetrates the sound waves produced from the jet engines of the aircraft. Similarly, an astronaut traveling in a spacecraft will observe the speed of light emitted from a source of light on the spacecraft, at a different speed of **c-v km/s** than the observer on Earth. Therefore, we believe that there is an error in the postulates of the STR. In Section 3.2, we shall review the concepts of GTR, first introduced by Einstein. The combined concepts of the special and general theories of relativity from Einstein will be referred to as E-theory.

3.2 Principles of the General Theory of Relativity from Einstein

In this section, we shall review the postulates of the general theory of relativity (GTR) as stipulated by E-theory. Einstein formulated the principles of GTR in 1918.

The principles of Einstein's general theory of relativity, based on the following postulates:

The principle of equivalence:

- The inertia of body is to be regarded no longer as an effect of absolute space but rather as one due to other bodies. In ordinary mechanics the motion of a heavy body (on which no electromagnetic or other force act) is determined by two causes:

    1. Its inertia tending to prevent acceleration with respect to absolute space;
    2. The gravitation of the remaining masses. The motion is determined by the distribution of the remaining masses in the universe.

- For events on the Earth, it states that all bodies fall at an equal speed under the influence of the Earth's gravity. For the motions of the heavenly bodies, it states that the acceleration is independent of the mass of the moving body.

- The laws of nature are represented by invariants for arbitrary transformations of the Gaussian coordinates, just as the geometric properties of a surface are invariant for arbitrary transformations of the curvilinear coordinates.

- Light waves arriving from distant stars are bent by force of gravity from a nearby star such as the sun. Also, the speed of light should decrease by the force of gravity from the sun. Einstein and several other scientists believe that light arriving from distant galaxies and nebulae are bent by force of nearby galaxies. This bending of light is believed to be the cause of multiple images observed under telescopes. The effect popularly described as the science of gravitational lensing, is extensively studied and the detailed mathematical model and analysis is being done by Dr. John Peacock at the Cambridge University. We believe that the multiple images of distant celestial galaxies are caused by the refraction of light from the Earth's atmosphere and the refraction of light caused by cloud of gaseous matter surrounds those distant galaxies.

Figure 3.1 Picture: Albert Einstein with Rabindranath Tagore from India.
Courtesy of the New York Times, August 10, 1930

In regard to the total energy content of matter, what is puzzling to us is that in the GTR, Einstein had talked about the effects of gravity on the motion of all objects in the universe. Yet, he did not account for the potential energy variations when he considered the total energy content of mass in the STR theory. He gave consideration to kinetic energy in his equations of the mass equivalence of energy. In his later years, Einstein did mention that his STR did not provide the correct results in accelerating systems and for the systems that do not conform to Hermann Minkowski's world lines. Long before Einstein formulated postulates of special and general theory of relativity, his tutor and mentor, Minkowski had introduced concept of coordinates of point p in the universe as comprised of four coordinates x, y, z and t. He stated that every point in the universe should be described as an event in space, which has a location and time associated with it. To describe motion of objects, and points in the universe among different inertial systems, he developed a system of world lines. For inertial systems that were moving at constant velocity with respect to each other, the world lines of points in those systems were rectilinear and followed rules of Euclidian geometry. For the systems which were accelerating, the world lines of points in the systems were curved and did not obey the rules of Euclidian geometry. In non-Euclidian geometry, the shortest distance between two points should be a curved line.

Before we describe new postulates of relativity, we like to add a historical note. When Einstein was in his Sixties, he realized some of the inadequacies in his theory of relativity. He understood that he could not figure out how to incorporate effect of gravity of large celestial objects on miniscule particles inside nucleus of atoms. Also, his theory of gravitation did not explain fundamental questions related to gravitation force among objects and at what speed the force propagates. In this book, we have answered these questions in much greater depth and with very high level of confidence than before. We like to point out that Einstein was disappointed to figure out when he introduced a cosmological constant in his field equation of general relativity. The presence of cosmological constant in his equation predicted a static universe. He thought, he missed a great opportunity by not describing continuous expansion or contraction of universe, later predicted by Hubble (see chapter 8). From this book's point of view, our universe is neither contracting nor expanding. It is infinite and open universe. Therefore introduction of cosmological constant in field equation does not harm and should not be a big deal. We do not agree with Hubble and Big-bang theory that our universe was created from point source of vast amount of energy. We think the greatest error of Einstein was in special theory of relativity that assumed speed of light, a constant.

Although many ideas presented by Einstein on theory of Relativity were highly speculative and controversial, he gained lot of respect and honor from scientific community in the world. He started norms and lead research that were not heard before. Outside of his career in scientific world, he was a great philosopher. In Figure 3.1 we have displayed Einstein's photograph with famous philosopher and poet from India, Rabindranath Tagore when he met him in India. The picture was retrieved from archive New York Times, issue August 10, 1930. In the next section, we shall state the principles of the special **Skylativity®** theory. Also, as we progress, we shall identify some of the main differences between the postulates of E-theory and K-theory. This should allow us to discuss the implications of the new theory to advance radio astronomy applications in the future.

3.3 Skylativity®: Sky's Theory of Special Relativity

In this section, we shall describe the postulates of the special **Skylativity®** theory, as stipulated by K-theory.

The principles of the special **Skylativity®** theory are as follows:

- Newton's Laws of classical mechanics are applicable to any frame of reference including quantum mechanics.

- The light is a wave. The speed of light is not constant **c**. It varies in various media such as the atmosphere, water, and glass, and depends on frame of reference. Also, it is different for observers in the relative state of motion with respect to light. The speed of light should be different for the observer who is traveling in space with the speed **v** as compared to other inertial systems in which the speed of light is **c** from his frame of reference. The speed of light should be related by the Galileo transformation [7], **c' = c + v** for the approaching scenario and **c' = c - v** for the receding scenario. The speed of light should be treated as a vector quantity. It should obey Newton's laws of motion for accelerating frames of reference.

The first systematic effort to indicate that the speed of light varies came from the Portuguese physicist, João Magueijo, in 1995; he was a research fellow in theoretical physics at Cambridge University. His work was published in a Discover cover story [16]. It is astonishing fact that he stated that a varying speed of light (VSL) could actually explain from where the cosmic unity (common basic matter on distant galaxies) of the universe comes. Also, he suspected What if black holes aren't really holes after all? As we shall discover later in Chapter 8, that the views of this book coincides with his view. Further, according to K-theory free space permeability constant μ_0 should not be treated as a constant value for a mobile frame of reference. It is well known fact that permeability constant depends on magnetic moment properties of constituent elements of matter. Also, frequency dependence of relative permeability of many magnetic materials has been studied extensively in modern times. Thus, we have proven that the speed of light should not remain constant but should vary in accordance with changes in μ_0 of the equation $\mathbf{c} = (\mu_0 \times \varepsilon_0)^{-1/2}$ **km/s**. More recently, the variations of ε_0 and μ_0 are being investigated by research professors at the University of Michigan in Ann Arbor, Michigan in designing very high frequency oscillator circuits.

- Whenever a material object attains the true absolute speed of light **c** (refer to Chapter 4 for the absolute speed **c**), with whatever cause or means, will reach an unstable condition. It shall radiate light, if the electrons in the constituent atoms attain an excited state and result in a transition event. When light is radiated, a portion of the kinetic and potential energy of the electrons involved in the transition event will be consumed. The light energy released is expressed by relation (4.13), as indicated in Chapter 4.

- The wavelength λ of emitted light is characteristic of the atomic structure of the matter. As a result of light emission, the energy level of particles will decrease as stated in equation (4.13). The velocity of the particle will decrease to some value below **c** to reach a stable state again. The resulting velocity of the particles can be computed from equation (4.12).

- In general, it is possible for a particle with finite rest mass to achieve the speed of light and above the speed of light **c** as long as the condition for the radiation event is avoided. The mass of the object should remain constant and should not increase with speed, as stated by Einstein's work, because the mass is the measure of the quantity of the matter in which it is contained. The upper limit on the speed **c** should be confined to the speed of the energy wave-type entities and should not be applicable to objects with real rest mass. In the future, it may be possible to define mass as a vector complex quantity with the real part known as rest mass and an imaginary part known as kinetic mass.

- The time-dilation should not occur for any frame of reference. It is always possible to compensate time dilation by design of clock in a free falling frame of reference to measure proper and absolute time. The proper and absolute time measurement will coincide with clock time measurement on static frame of reference. This implies that time; length and mass measurements should remain the same and should have values independent of frame of reference. The Lorentz transformation for measurement of time, length and mass is not required.

- The final and most interesting postulates from **Skylativity®** theory is that the mass of any light/heat or electromagnetic energy waves radiating substance is constant. In any electromagnetic wave radiation event no mass is consumed. We shall prove this fact that total number of particle counts in the element that constitute radiation event before the chemical/nuclear process and after the process completion is unchanged. When we state the particle count, it means that if we add a number of protons, neutrons and electrons of the substances that caused a process, and radiated light or other forms of energy waves, the number will be the same at the end of the reaction.

  For instance, when we burn coal, if we add particle counts of carbon and oxygen gas before the burning process and compare it with particle counts of carbon dioxide gas generated from the process the count should balance on both sides of the equation:

  $$\mathbf{C + O_2 \Rightarrow CO_2}$$

  Mathematically, if $\mathbf{M_1}$ to $\mathbf{M_n}$ correspond to the masses of elements $\mathbf{X_1}$ to $\mathbf{X_n}$ before the reaction and $\mathbf{M_1}$ to $\mathbf{M_k}$ are the masses of elements $\mathbf{Y_1}$ to $\mathbf{Y_k}$ after the reaction, then the total mass:

  $$\mathbf{M_1 + M_2 + \ldots\ldots + M_n = M_1 + M_2 + \ldots\ldots\ldots\ldots + M_k}$$

  for any value of **n** and **k**, no matter how many watts of the waveform energy is released as a result of reaction. We shall suggest an experimental setup comprising of Edison's bulb to prove this fact in Chapter 4

- The light wave energy released from the thermonuclear burning process of hydrogen into helium is computed from the relation:

  $$\mathbf{E = m \times c^2}$$

  where **m** is the consumed mass of hydrogen and **c** is the true speed of light.

  Since the application of the Einstein equation is specific to the thermonuclear fusion process, and it does not quantitatively state how much of the mass is transformed into what types of energies, infrared, visible, X-rays, and ultra violet rays, the value of the formulations is diminished after this analysis. Further, only the electron mass (a very small fraction, 0.7% of the hydrogen mass) is transformed into energy within the sun.

In any other nuclear reaction, Static mass of radio-active material is not transformed into light waves. According to Baryon number rule the total number of protons and neutrons in the universe remains constant. In a radioactive event total number of protons and neutrons before the reaction and after the reaction should not change. The matter constituent elements change such as from $^{238}\mathbf{U}$ to $^{235}\mathbf{U}$ etc. because atomic configuration is altered. The vast amount of binding energy is released in fission nuclear reaction. In the next section, we shall state the principles of the general **Skylativity®** theory.

3.4 Skylativity®: Sky's Theory of General Relativity

In this section, we shall describe the postulates of the special **Skylativity®** theory, as stipulated by K-theory. It will be evident, that we agree on some of the postulates of the general theory of relativity, as stated by Einstein. We shall explain the differences between the general theories of relativity from E-theory and K-theory.

The principles of the general **Skylativity®** theory are as follows:

<u>The principle of equivalence</u>

- The inertia of the body is to be regarded no longer as an effect of absolute space but rather as one due to other bodies. In ordinary mechanics the motion of a heavy body (on which no electromagnetic or other force act) is determined by two causes:

  1. Its inertia tending to prevent acceleration with respect to absolute space;
  2. The forces of gravitation from the remaining masses. The motion is determined by the distribution of the remaining masses in the universe.

- For events on the Earth, it states that all bodies fall at an equal speed under the influence of the Earth's gravity. For the motions of heavenly bodies, it states that the acceleration is independent of the mass of the moving body.

- The laws of nature are represented by invariants for the arbitrary transformations of the Gaussian coordinates, just as the geometric properties of a surface are invariant for the arbitrary transformations of the curvilinear coordinates.

- The light waves should not bend by the force of gravity, as predicted by Einstein and others. Also, the speed of light should not decrease by the force of gravity. The light energy waves do not have a center of mass that is static because the waves are propagating. For the force of gravity to act and deflect the position of another entity, it requires that the entity should have a static center of gravity at an instant in time in its own frame of reference. Therefore, the light waves from the distant galaxies should not be deflected by the force of gravity from nearby galaxies and stars. The lack of the bending of light should result in the lack of multiple images of the distant galaxies, as seen through a telescope on Earth. As explained, the refraction of light from the Earth's atmosphere and the passage of light from the gaseous clouds surrounding the galaxies, are the main reasons for the multiple images of the galaxies. If the gravitational bending of light was the cause of multiple images, too many celestial objects would have multiple images observed in a telescope in an observatory. It has been discovered that only a few celestial objects show the multiple images. This substantiates our argument that the light does not bend by the force of gravity.

- Gravity waves and field does not exist. The force of gravity occurs when two objects with non-zero rest mass are in proximity. Further, multiple strength gravity field lines could co-exist when several objects are in vicinity. Therefore, unique solution for field equation does not make sense. Loss of mass caused by radiation of strong gravity field and waves from vast celestial objects and binary systems is not correct.

- The speed of the propagation of the force of gravity is a universal constant and is the same as absolute true speed of the propagation of light waves **c**. We shall describe the consequences of the finite speed of propagation of the force of gravity in viewing the range of objects in the universe and their relationship with the prediction of the positions of various celestial objects in real time.

In regard to the constancy of **c** in his later work, Einstein agreed that the upper limit on the speed of light **c** is valid only for Minkowski's world geometry, in which the acceleration **g** has no value [7]. In the presence of the acceleration **g** light, the speed may exceed **c**. For stars, $\mathbf{V = 2\pi\, r/T}$ if the distance of the star **r** is $> \mathbf{1/(2\,\pi \times 365)}$ light year, the speed should be higher than **c**. The period **T** is one day for the Earth. Further, the GTR predicts that, in the arbitrary Gaussian coordinate system, not only does the velocity of the light change, but the light rays bend. The world lines of light are geodesic lines, similar to the inertial orbits of the material bodies, and hence, like the matter, will in general be curved [5]. As per our explanation in Section 1.2, the trajectory of light should not bend or be deflected by gravitation. The light is a wave and it may have energy. The main difference between mass and energy is that mass has a center of gravity and finite volume. When the light propagates in space, the space contains energy. The energy in space lacks volume because the energy is defined as the work required in changing the Newtonian state of the object. This implies that the world lines of light in space do not curve according to this work.

In our universe, which is comprised of a very dilute system of stars, as described by Dutch astronomer, Willem de Sitter, it is highly improbable that light or any electromagnetic signal radiated from the Earth will be received without any change. If light waves should bend as predicted by Einstein and others then light wave signals transmitted from the Earth should return without reflection. In the event non-reflected light is received by the Earth because of curvature of space the received light should have a zero Z-shift. To date there is no record of event where transmitted light is received back with zero red shift. Further, Einstein stated that space will warp curve and word line of light rays from a source will close on itself. Our belief is that space is endless and has no region or surface boundary. Therefore, light rays will never return unless it is reflected by a heavenly objects and it will not close on itself. We disagree with the GTR on the issue of the bending of light in presence of strong gravitational fields. As stated earlier, light wave should not bend because it lacks center of gravity. Also, life time of one wave of light is very small to have any effect of gravity. We have cited several primary differences in the postulates of STR from Einstein and this work. This effort points out that there is a definite need for revisiting the Lorentz transformation for two inertial systems moving with different speeds relative to each other. Also, the postulates of the general and special theories of relativity from Einstein should be updated to account for the VSL as suggested by Magueijo

One of the side effects of the limitation on the speed of energy wave-type entities, such as light waves and electromagnetic waves, is that it imposes a limit on the speed at which the information may be communicated over any channel. Therefore, even if one could travel at or above the speed of light with a vehicle, there is no means to send or receive the information from the vehicle above the true speed of light **c**. This is a mystery that has to be solved before one can attempt to travel above the speed of light because the remote control of these vehicles is a challenge, a problem to be solved. In October 2009, we had a meeting with a well known physicist in India, Dr. J. J. Rawal, President, The Indian Planetary Society. The meeting was held in his office in Gorai, Borivali, and a suburb of Mumbai. We discussed in great detail, many issues that arose from the new postulates of **Skylativity®**. Our discussions with him and another leading physicist, Viren Dave, were highly useful. After very careful consideration on the ideas presented, we concluded that physics experts in the world should take a fresh look at the earlier theoretical physics ideas of relativity, introduced a century ago. As per the postulates of the **Skylativity®** theory and the concepts introduced by Magueijo, the speed of light is variable and should depend on the frame of reference. Therefore, in the next section, we shall discuss the modifications required in Maxwell's field equations, to account for the changes in the speed of light among the different references.

3.5 Maxwell's Field Equation for Varying Speed of Light

Before we can discuss the modifications required to Maxwell's field equations, for variations in the speed of light as the speed of reference varies, let us restate the equations in their original form, as well as differential forms. First, we will state Maxwell's four equations in integral form. Next, we shall state them in differential form because it is simpler to describe the effect of light speed variation.

Maxwell's equation in integral form:

$\oint \boldsymbol{E} \cdot \boldsymbol{dA} = \mathbf{q/\varepsilon 0}$ Gauss's law **(3.1)**

$\oint \mathbf{B \bullet dA = 0}$ Gauss's law in magnetism **(3.2)**

$\oint \mathbf{E \bullet ds = - d\Phi_B/dt}$ Faraday's law **(3.3)**

$\oint \mathbf{B \bullet ds = \mu_0 I + \varepsilon_0\ \mu_0 d\Phi_E/dt}$ Ampère-Maxwell law **(3.4)**

Maxwell's equation in differential form:

$\mathbf{\nabla \cdot E = 4\pi\rho}$ **(3.5)**

$\mathbf{\nabla \times B - 1/c(\partial E/\partial t) = (4\pi/c)\ j}$ **(3.6)**

$\mathbf{\nabla \cdot B = 0}$ **(3.7)**

$\mathbf{\nabla \times E + 1/c(\partial B/\partial t) = 0}$ **(3.8)**

After carefully looking at these equations in the integral form, one may discover that three of the four equations Carl Friedrich Gauss's law, Gauss's law in magnetism, and Faraday's law are not affected by the variation in the speed of light. However, one may observe that the André-Marie Ampère and James Clerk Maxwell (Ampère-Maxwell) law's fourth equation needs to be modified. It will be obvious from the fourth equation stated in the integral form; the multiplication factor μ_0 involved in both terms on right hand side should be changed to μ_r. The reason for this is a change of electric flux through any surface bounded by a moving path creates a changing magnetic flux. We are suggesting this change because a free space in another frame of reference exhibits different permeability characteristics and it is reflected in the differences in the speed of light for that frame of reference. Also, the permittivity of free space would be different if the free space is moving with respect to a stationary frame.

The effect of the variation in the speed of light modifies the computations. Conventionally, the speed of the electromagnetic waves is determined by the expression:

$$\mathbf{c = (\mu_0 \times \varepsilon_0)^{1/2}\ m/s} \quad \mathbf{(3.9)}$$

where μ_0 is permeability constant and ε_0 is permittivity constant for free space (vacuum) which is stationary with respect to the observer

by substituting a value of: $\mu_0 = 4\pi \times 10^{-7}\ NA^{-2}$ and
$\varepsilon_0 = 8.854 \times 10^{-12}\ Coulomb^2/(newton\ meter^2)$

we get: $\mathbf{c = 2.99792458 \times 10^8\ m/s}$ a universally accepted constant

The new form of the equations suggested by us, is: $\mathbf{c = (\mu_r \times \varepsilon_r)^{1/2}\ m/s}$. Therefore, in a generalized frame of reference, the speed of travelling the electromagnetic waves is computed by the relation:

$$\mathbf{c = (\mu_r \times \varepsilon_r)^{1/2}\ m/s} \quad \mathbf{(3.10)}$$

where μ_r is permeability constant and ε_r is permittivity constant for moving frame of reference respectively

The new constants (μ_r **and** ε_r) affect the speed of light similar to the effect of (μ_r **and** ε_r) in determination of mutual inductance and coupling capacitance in electric circuits reactance computations for circuit elements designed with more than one material for coiled inductors and dielectric capacitors. The symbols μ_r and ε_r, relative permeability and permittivity, are modified constants to account for the changes in the properties of space because it moves in relation to the observer. In the next section, we shall explore the effects of light speed variations on the solution of Einstein's field equation.

3.6 Effect of New Postulates on Einstein's Field Equation

Since the exact description of Einstein's field equations require rigorous mathematics, which is beyond the scope of this book, we shall only touch on the changes that are introduced by the variations of the speed of light on the equation. First we shall state Einstein's field equation in its present form. It is important to note that although the equation was discovered by Einstein, it was also discovered independently by David Hilbert who deserves to share credit because he published the field equations a few days earlier than Einstein. The versatility of Einstein's field equation comes from the fact that it gives state of gravitational field surrounding a mass distribution. For instance, equations for the gravitational field surrounding the sun or the gravitational field surrounding a black hole could be derived from his field equation. The nonlinear field equation to describe gravitational field is [19].

$$R_{\mu\nu} - 1/2\, g_{\mu\nu}R = -\,(8\pi G/c^4)\, T_{\mu\nu} \qquad (3.11)$$

The exact nonlinear equations are implied by the linear equations is a remarkable feature of Einstein's theory. However, this tight connection between the exact equations and the linear approximation no longer exist because according to **Skylativity®** theory **c** cannot be treated as a constant that violates the principle of general invariance. Thus, the feature of **Skylativity®**, that the speed of light depends on the frame of reference, decreases the importance of Einstein's field equation, as it destroys the simplicity of handling. The primary advantage of solution of Einstein's field equation was for point particles in local geodesic coordinates, the equation of motion of the particle showed that the energy and momentum of the particle do not change. Since in Einstein's theory all particles obey the geodesic equations of motion for any arbitrary system, both Galileo Galilei's principle of equivalence and Newton's principle of equivalence were satisfied. As a result of this it was established that gravitational mass equals inertial mass.

Several applications of Einstein's equation to solve the deflection, retardation and the red-shift of light in the sun's gravitational field are discussed with analytical treatment in great detail [19]. From our book's viewpoint, the predominant cause for the bending of light as it passes by the strong gravitational field of the sun, is not the force of gravity, but may be the refraction of light rays as they pass through the gaseous atmosphere near the sun and another refraction of light through the Earth's atmosphere. From our computations in Section 1.2, you will be convinced that our statement makes sense. Therefore, a detailed mathematical analysis of the bending of light is a misconception and is over killed effort. The most admirable fact as regards to Einstein's gravitational field equation it solves gravitational field of an arbitrary complex object fabricated by carving any shape within the finite universe [20]. In its entirety, Einstein's field equations are based on Tensor and Spinor Algebra. It was a great discovery when he came up with these remarkable ideas in the early-20th century.

It is important to understand the limits of Einstein's field equations. One of the issues is that in his regime, Space and time are curved by the gravitational field of a geometrical shape of a surface. However, the extent of this curvature is expressed by shape of the surface much more so than the property of the matter from which the surface is fabricated. Therefore, his equations do not provide solution of gravitational field for geometries enclosed by arbitrarily complex surfaces in the infinitely large, open universe. His equation pauses a computationally NP incomplete problem for the case of random shaped geometrical shapes in the open, infinite universe. This is especially true for gravitational fields which have a finite speed of propagation. Another setback happened to Einstein's formulation in the 21st century when physicist discovered that a huge percentage of fast moving neutrinos from the sun and arriving from other stars in space are passes through the Earth's core in a straight line without interaction. It is well known that Einstein's field equations relies heavily on rules of non-Euclidean geometry in which the shortest distance between two points on a surface with curved topology is not a straight-line. Because neutrino may penetrate the Earth's core, it is safe to mention that no volume in space within the universe could be carved that will follow the rules of non-Euclidean topology in the strictest sense. The topologies that obey the rules of Euclidean cause severe restrictions on the solution of Einstein's gravitational field equations.

In another application, though Schwarzschild's solution of Einstein's equation claims that it provides the exact solution for the perihelion precession of planetary orbits in the field of the sun and for the gyroscope precession effect, we think the pain is not worth the effort. Our conclusion is based on fact that the difference between orbital equation from Newtonian theory and from Schwarzschild field equation, the relativistic correction to the motion is extremely small [19]. An excellent analysis and data for the perihelion precession of all the planets of the solar system is provided by Hans Ohanian and Remo Ruffini.

3.7 Summary

In this chapter we started with the discussion about the principles of the special and general theories from Einstein. To refresh reader's memory we stated postulates of STR and GTR in sections 3.1 and 3.2. During our discussion we pointed out that why the speed of light should not be constant and independent of frame of reference. Further, we proved that force of gravity should not have any effect on multiply polarized beam of light. Both fields created by electric charges, as well as magnetic poles seem to have no effect on trajectory of light. Therefore, force of gravity should not bend light waves. In Sections 3.3 and 3.4, we described the postulates of the special and general **Skylativity®** theories. We established the fact that the speed of light is variable and obeys laws Galileo's transformation for the speed of computations among different inertial systems and does not obey laws of the Lorentz transformation as stated by Einstein. Also, we stated that upper limit on the speed of light is confined to energy wave entities and not to the particles. The speed of the particle can be increased above the speed of light, if the particle does not start light radiation. If the particle starts a radiation event, it will result in the loss of kinetic energy that will be reflected in a decrease of speed. Because the speed of the wave entities have an upper limit that equals the absolute speed of light **c**, it has a side effect that the information transfer cannot occur at the speed beyond **c**. These limitations cause severe problems for the remote control of vehicles that travel at or above the speed **c**. Further, this limit requires that the data from sensors of position and velocity of objects should be available in advance to apply control signals to the systems even if the systems are local.

In Section 3.5, we discussed the differences between the original form of Maxwell's field equations and the equations suggested by us. Specifically, we showed that the Ampère-Maxwell law equation needs to be updated to reflect the different values μ_r, ε_r of permeability and permittivity in moving free space frame vs. stationary frame of reference values μ_0 and ε_0. In the following section, we described the modifications required in solutions of Einstein's field equation. We concluded that space time curvature predicted from solution of Einstein's gravitational field equation for random shaped geometrical surfaces are functions of the shapes of surfaces more than the property of the matter from which the surfaces are fabricated. Because the fast moving tiny particles, neutrinos, can penetrate any volume bound by any shaped surface, it is difficult to construct a surface that is truly of a non-Euclidean nature. Therefore, the solution of Einstein's field equation does not provide the correct solution for the gravitational field in real time as it is an NP incomplete problem for non-Euclidean topology surfaces.

In the next chapter, we shall show the limitations of Einstein's theory of relativity by explaining the physics behind the creation of light in a common incandescent lamp originally invented by Edison. Then we shall prove that the Lorentz transformation and the concepts of time dilatation are useless in determining the relationship of the time space coordinate among different inertial systems.

4.

Limitations of Einstein's Theory of Relativity

In this chapter we shall investigate the physical applications that completely contradict the notions postulated by Einstein. Specifically, in Section 4.1, we will analyze physics behind Edison's bulb, a common incandescent bulb to discover where Einstein's theory of relativity failed which state that infinite supply of energy is needed to accelerate the particles at rest to the speed of light. Also, in the Section 4.2, we shall provide an experimental proof of the fact that when light energy waves are emitted from bulb, there is no loss of mass from the tungsten filament from Edison's bulb. The main purpose of this experiment is to show that there is no mass to energy conversion when light energy is released in a thermal radiation event. The observation made from the results of this experiment may be extended to nuclear radiation events.

In Section 4.3, we will illustrate that the speed of light **c** in two inertial systems **S** and **S'** moving with respect each other with a non-zero velocity vector difference cannot be the same in amplitude value and phase value. We derive this fact by applying the principle of the laws of conservation of the energy among different inertial systems. Also, it made sense to describe the mass of the object in the moving frame of reference as vector quantity. Therefore, in Section 4.4, we explained that total mass of any object in all frames of reference have static component and an orthogonal dynamic imaginary component. In section 4.5, we explained theory of gravitation from Einstein. We showed that the theory faces severe limitation in realm of quantum particles. In Section 4.6, we shall prove that concept of time-dilation introduced by Einstein and value of time dilation as per computations suggested by the Lorentz transformation does not make sense. We are stating this fact because it is always possible to re-design clocks such that the measured time will display the same value for both inertial systems moving with respect to each other. Therefore, absolute unit of time is not alterable. In the section 4.7, we will show that multiple images of quasars and distant galaxies in our universe are not caused by the bending of light due to force of gravity of intermediate stars on light from those objects, a Gravitational Lensing projected by several western astrophysicist [4]. Instead we will show that multiple images are caused by the refraction of light arriving from those celestial objects when they are passing through the Earth's atmosphere.

As stated earlier, in Section 4.1, we shall begin our discussion on Edison's bulb and show how a common incandescent light bulb can disprove the major concepts of the special theory of relativity (STR), originally introduced by Einstein.

4.1 Edison's Light Bulb

In this section, we shall examine the light emission process for a bulb lamp. Our detailed analysis will prove that the principles of operation of such a simple apparatus are in conflict with the fundamental concepts of the STR. In particular, our description shall prove that with a finite, small supply of electrical energy to the bulb transforms into light energy waves that were at zero speed before the application of the electric current and power to the bulb.

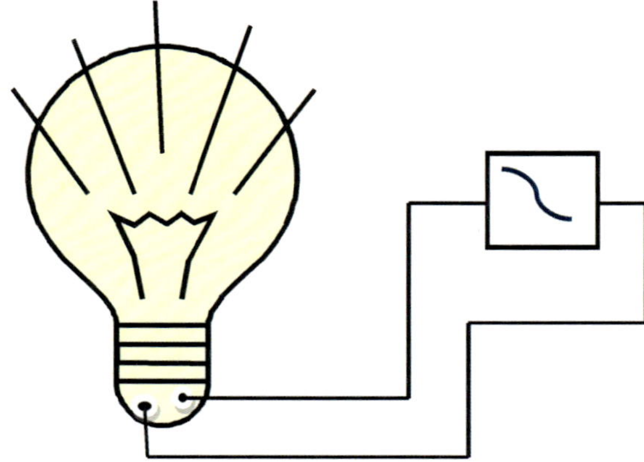

Figure 4.1 Simple incandescent light bulb. Operation of a light bulb indicates that the speed of light is achieved by a finite amount of energy.

Let us consider the example of light emission from a common incandescent light bulb first invented by Edison in 1879. He used carbonized cotton thread material for the filament wire in his bulb [39]. In modern times, the filament of a light bulb is formed from Tungsten wire which is highly resistive material. When an electric current flows through the filament, intense heat is produced due to the resistance of the wire. The atoms of the element of the wire are heated at a very high temperature. Subsequently, the filament inside the bulb glows and emits light. Microscopically, the high temperature increases the vibration inertia and the molecular speed of the filament. The atomic binding energy still keeps the molecular structure of the constituent of the filament element intact. The free electrons, of the constituent atoms of molecules of wire on the surface, are excited and trigger the electron orbit transition event. Therefore, the filament starts radiating light.

We would like to examine the situation more closely. When the current is off in the filament circuit, there is no light from bulb. It is dark because the photons as claimed by proponents of particle theory resident on filament are at zero frequency and are at rest. When the current is switched on, the light bulb emits light. This implies that photons on the wire are accelerated to the speed of light from rest (zero initial speed) with the application of finite energy. This further suggests that the photon particles have gained the speed of **c** and their frequency is changed to that of visible light in an infinitely small time. As per the postulates of the STR, particles cannot be accelerated to the speed of light. It would require an infinite amount of energy because the particles' relativistic mass increases with speed and becomes infinite when the speed of light is attained.

The radiation of light from Edison's bulb illustrates that, Relativity theory fails. Also, this example proves that the light is a wave because only waves can change the frequency and speed in infinitely small or zero time. In practice, eventually, the Tungsten wire is oxidized due to residue air and leakage of air molecules through pores of glass casing of the bulb. Therefore, the bulb ceases to radiate light when all Tungsten in the wire is depleted or oxidized and electric current flow is prohibited due to the mechanical breakdown of wire. We are surprised to notice that Edison's invention preceded Einstein's work by almost 20 years yet Einstein did not realize that Edison's lamp is an example where his theory failed. Light emission from more recent inventions of mercury vapor lamp, gas molecules sodium lamp signs and various neon lamp signs [39] and Edison's bulb further validates new ideas proposed in this book. When high voltage is applied between anodes and cathodes of low pressure (almost vacuum) gas lamps current conduction occurs because electric circuit is completed by passage of current through arc. The light is emitted from orbiting electrons as they jump to a lower quantum state for the extreme high temperatures reached by the thermally agitated molecules.

In the next section, we shall describe the details of a simple experiment that allows us to prove that the mass of light and the thermal radiating substance, such as the Tungsten wire filament of a bulb, remains unchanged after several hours of operation radiating the light energy waves continuously.

4.2 Mass of Tungsten Filament in Edison's Bulb

We are proposing a simple experiment to establish a fact that when light energy is radiated from a substance no quantity of matter is consumed. For instance, for the bulb shown in Figure 4.1, if we measure the mass of the Tungsten filament after several hours of operation, the mass should not be affected. For this purpose we shall perform a simple test. We shall measure mass of filament when the bulb was never lit at time zero. We shall designate mass of wire as $\mathbf{M_0}$ **g**. Next, we shall operate the bulb continuously for 100 h, 200 h, 500 h, 1000 h, and 10000 h in succession. After each hours of operation we shall measure the mass of the Tungsten wire and we shall designate the values of masses as $\mathbf{M_{100}}$, $\mathbf{M_{200}}$, $\mathbf{M_{500}}$, $\mathbf{M_{1000}}$, and $\mathbf{M_{10000}}$. We shall discover that various measured mass values shall not differ to the slightest extent if we prevent any oxidization of the wire from the initially measured value of the wire mass.

In the expression form:

$$\mathbf{M_0 = M_{100} = M_{200} = M_{500} = M_{1000} = M_{10000}} \qquad \mathbf{(4.1)}$$

In the next section, we shall focus on the computations of the total energy content of the same objects in different inertial systems. We will apply the principle of the conservation of energy to prove that if the material mass of the matter in the systems moving with respect to each other is different, the speed of light in both systems cannot be the same.

4.3 Energy and Mass Paradox for Matter

According to the STR, the total energy content of matter in a reference system **S** at rest is expressed by the relation:

$$\mathbf{E = M_0 \times c^2}$$

where $\mathbf{M_0}$ is the mass of matter in system **S**
which is at rest, and **c** is the speed of light

The mass of matter $\mathbf{M_v}$ in a different inertial system **S'** which is moving with velocity **v** with respect to **S** is related to the reference mass by the Lorentz transformation according to current theory. It is expressed by the relation:

$$\mathbf{M_v = M_0 / (1 - v^2/c^2)^{1/2}} \qquad \textbf{(4.2)}$$

For a detailed discussion, you may refer to the equation (78) from page 273 in [7]. Therefore, energy content of the matter when measured in system S' should be $\mathbf{E' = M_v c^2}$
However, the total energy content of the matter should not change between **S** and **S'**. The mass or quantity of matter contained in an object has not changed even though the speed is changed in two systems. Therefore, ability of mass to be transformed into final form of energy light is not altered. It is established that the total energy of any system is conserved in the universe at any time. Hence:

$$\mathbf{E' = E \text{ or } M_0 c^2 = M_v c^2}$$

From this equation, $\mathbf{M_0}$ and $\mathbf{M_v}$ are different, which implies that the speed of light **c** cannot be the same in both system **S** and **S'**. Therefore, our conclusion is that **c** should not be constant and the value should not be equal for both systems.

In fact, according to the Lorentz transformation, the measured length l in system **S'** is related to and shorter than $\mathbf{l_0}$ in systems **S** by:

$$\mathbf{l = l_0 (1 - v^2/c^2)^{1/2}} \qquad \textbf{(4.3)}$$

and measured time **T** in system **S'** is related to and longer than $\mathbf{T_0}$ in systems **S** by:

$$\mathbf{T = T_0 / (1 - v^2/c^2)^{1/2}} \qquad \textbf{(4.4)}$$

The speed, which is the ratio of distance over time, should be proportional to **l/T**, which from equations (4.4) is:

$$\mathbf{l_0 (1-v^2/c^2)/ T_0} \qquad \mathbf{(4.5)}$$

Therefore, the velocity of light should be different in the two inertial systems.

In reference [7] page 266, it was proven that the speed of light is constant and independent of frame of reference. The proof is invalid because it was derived from velocity equations of the Lorentz transformation for two systems **S** and **S'**. However, the transformation equations were based on the assumption that the speed of light is constant. Therefore, the proof is not sufficient. If you do not assume the constancy of the speed of light between systems **S** and **S'**, the time dilatation and the measured length of rod should be adversely effected. Our expectation would be that the measured length will be shorter and the measured time will be longer than the case of the constant speed of light. Clearly, one needs to derive the Lorentz transformation again, with the assumption that the speed of light should be different in system **S'** from the normal value **c**.

In the next section, we shall propose a new way to describe effective mass in a mobile system. We shall stipulate that mass of objects in the universe is a vector quantity similar to velocity and acceleration. We shall explain that it made more sense to split and represent total mass of moving objects into two components static mass and dynamic energy mass.

4.4 Relativistic Mass a Complex Quantity

As stated in Section 4.3, we believe that mass of any object in motion comprised of two components. An eternal or rest mass which is a component of the mass that is real. What we mean by real quantity is the mass component which contributes to influence position of center of gravity of the object. We define eternal or rest mass of an object is that mass of the object which is measured when the distance between measuring apparatus and the object does not alter during the period it is measured. The other component of the mass is due to mobility of the object. This mass represents energy mass component an imaginary component of mass. We suggest this mass is directional component which moves the center of gravity of object to a new position at time δt after time t_0. Therefore, in relativity, mass should be represented by a phasor. Symbolically mass of objects in motion should be represented by:

$$\mathbf{M = M_r + jM_e} \qquad \mathbf{(4.6)}$$

where $\mathbf{M_r}$ is stationary mass and $\mathbf{M_e}$ is energy or dynamic mass

This complex phasor representation of mass provides a convenient means to handle the mass of objects in different inertial systems. The main advantage is that it avoids the singularity associated with relativistic mass representation by the Lorentz transformation formulae. Also, we infer that there is no increase in real or rest mass of object when it is accelerated by any means. Therefore, there is no upper limit on the speed of travel of the objects with real mass. The mass of the objects travelling near the speed of light does not approach infinity, as per the predictions made by Skylativity® theory. Further, the complex representation of mass allows solving the differential response equations of mechanical systems analogous to the solution of differential response equations of electrical circuits with impedance notion. We shall explain the importance of impedance notation for mass in Chapter 8 when we shall understand the meaning of mass of fundamental particles.

In the next section, we will discuss the ideas of Einstein on the theory of gravitation, also known as the general theory of relativity (GTR). The theory is widely regarded as an intellectual triumph of 20th century science. It is claimed that mathematically, it is beautiful and in observation it has withstood the most stringent tests. Einstein had woven the gravitational field into the fabric of space and time. He suggested that gravitational force stands apart from all other forces because it manifests curved space time geometry. According to theory of gravitation, matter through gravity conveys how space-time should bend and curved space-time tells matter how to move. As we shall see that his theory is only applicable for large celestial objects. For quantum dimensions and particles force of gravity is much weaker than strong charge forces. Further, theory of gravitation is unable to explain why space-time without matter and undefined shape may curve by gravitational field of dense objects.

4.5 Theory of Gravitation

The gravitation theory introduced first by Einstein, in the 1920s, was called the general theory of relativity (GTR) to distinguish it from the special theory of relativity (STR) that does not account for gravity effects on motion of light in the presence of heavy celestial objects. The essential feature of the theory is that the universe is full of matter, and matter warp space-time by gravitational force created by the mass of matter in such a way that all bodies fall together. According to conventional wisdom of physicists, any theory that is accepted needs to be confirmed by experiments and verified by results observed. This theory made three major bold predictions that depended on a massive object like the sun to provide tests for experimental verification and defense.

1. The gravitational bending of rays of light from a star by the sun.
2. The advance of perihelion of the planet Mercury
3. The gravitational red-shift of the solar spectrum.

From our point of view results of these tests are not sufficient to prove the axioms. In particular, Einstein never clarified if gravity should affect a mass-less entity light waves in the same way as object with masses in test number one. We claim that his assumption, light without any rest mass will deflect by force of gravity is not correct. As per our discussion in Section 1.3, in Figure 1.4, the light waves coming from a star behind the sun are more likely to be deflected by the change in the index of the refraction of the atmosphere surrounding the Earth, and also the rays are refracted by the hydrogen gas surrounding the sun. Our reasoning is that rays of light arriving from the star behind the sun do not have a mass and static position of center of gravity. For force of gravity to act from one large body to another it is imperative that both bodies should have center of gravity. When we discussed this viewpoint with physicist and Emeritus Professor Dr. Jayant Vishnu Narlikar during our recent visit to India at his office IUCAA in Pune, he disagreed. He defended Einstein's position by stating that space in the vicinity of high gravitational field warps. Therefore, the trajectory of light is curved by gravitational field of dense celestial objects. Again we claim that because space lacks any matter there is no surface or boundary may be associated with a fabric in space. Hence, space should not warp or distort which does not have center of gravity or mass. Our argument is consistent with the magnetic force field generated by the North and South Poles on Earth. The magnetic field only affects material that has magnetic and polarization sensitivity. Therefore, in our opinion, the results of the measurement of the bending of light from a star during a total solar eclipse are not a good test. The GTR claims that it has brilliantly explained gravity but no where it answers the question, what is the velocity of the gravitation force, if any? The speed of the force of gravitation should play a very important role in the determination of the bending of rays of light without mass; those are propagating at the speed of **300,000 km/s**.

As far as the second test is concerned, the advances of the perihelion of the planet Mercury, we believe that the position of planet Mercury shifts by **43 arc s** per century. This shift is caused by the reduction in mass of the sun as it radiates the tremendous amount of solar energy per second, based on the Newtonian gravitation theory. Therefore, the result of this test does not conclude that theory of gravitation from Einstein is correct. For the third test, the gravitational red shift of the solar spectrum, the shift is miniscule even for a large object such as the sun. The measured shift amounts to **14 m A^0** which is very difficult to observe to form a solid conclusion of the test. From this discussion it is evident that the experiments performed to prove theory of gravitation from Einstein were not stringent enough to prove his points. Nevertheless, we shall continue our discussion about the theory of gravitation and see how he arrived at the concepts of the different scales for mass, length, and time for the measurements performed in systems that are moving with respect to each other at a constant velocity, while maintaining a constant speed for light.

It is widespread opinion that the theory of gravitation, as proposed by Einstein a century ago, was difficult to comprehend when he first formulated the theory. Unfortunately, he could not use the concepts of the standard model, and its interacting forces, that are now familiar, and may make the approach and theory much simpler. Einstein asked simple questions to model his theory of relativistic gravitation. First of all it must confront the fact specific to gravitation discovered by Galileo all body must fall equally fast regardless of their size, shape or composition in gravitational field of a large massive object. Secondly the theory should be universal that is all material bodies are involved in this interaction of gravitation.

Lastly, the theory must fit within the electromagnetic equations of Maxwell and with the STR. As we shall see later, that though the theory of gravitation claims to explain the bending of light by the sun, we have given reasons and are not completely satisfied with the explanation. Further, we discussed in Chapter 3 that Maxwell's equations need to be updated to account for the variations in the speed of light among different frames of reference. To explain the concepts of Einstein's theory of gravitation, he employed a simple illustration from force of gravity on bucket filled with atoms. Let us look at this example and see if it made sense. The details of his experiment are as follows.

Imagine a tower on the Earth carrying an endless chain of buckets, running over a wheel on the top and another at the bottom, as illustrated by the arrangement in Figure 4.2. Now, fill every bucket with the same number of atoms of the same elements. However, we made sure that atoms on side **E** are in excited state and atoms on side **G** are in ground state. According to the STR, atoms in an excited state are heavy; though we disagree with this fact, let us carry on the conversation. Therefore, buckets on **E** side will move downward. As a result buckets on **G** side will move upward. Now, stimulate the atoms of bucket at the bottom to emit radiation in our hypothetical experiment. The atom in the excited state in the bucket will transition to ground state. If we focus the emitted light from the bottom bucket on **E** side to the top bucket of **G-side**, by quantum theory this light should have just the right frequency to be absorbed by the atom to put it into an excited state **E**. A miracle? We have produced an ideal source of energy without external input of energy. According to principles of thermodynamics this is impossible unless the buckets on both **E** and **G** sides are at the same height. So, where is the discrepancy in the principle of operation of the bucket experiment? Let us explain the reasoning behind that next.

The inability of atoms to absorb the light of a given frequency at a different level, prompted Einstein to incorrectly conclude that the frequency of the light emitted at some level, should be less (different) if measured at a higher level than if it is measured at the level where the emission occurred. From our point of view, the atoms at the different level will require a shifted frequency to be absorbed because the energy state of the quantum particle at a different height is different than at the same height. This fact suggest that if by some means energy state of quantum particle is restored to the same value as the one at the same level it would absorb light of frequency of origin. At the present state of technology, this measurement is not impossible. Of course, for this scenario, in this the experiment it is difficult to replicate the measurement exactly. However, the clocks employed to measure time at the different elevations, may be accelerated or retarded to alleviate the effect of the difference in the gravitational potential energy, as seen by the atoms inside a Cesium clock. As we shall discover, an incorrect conclusion drawn by Einstein has adversely impacted the science community at large to a greater extent than perceived by many physicists of past and present generations. Fortunately for Einstein, he stated that the effect of such acceleration may not be distinguished from the effect of gravitational force in the frame of reference under consideration. We shall examine further, the consequence of the incorrect observation in Section 4.6.

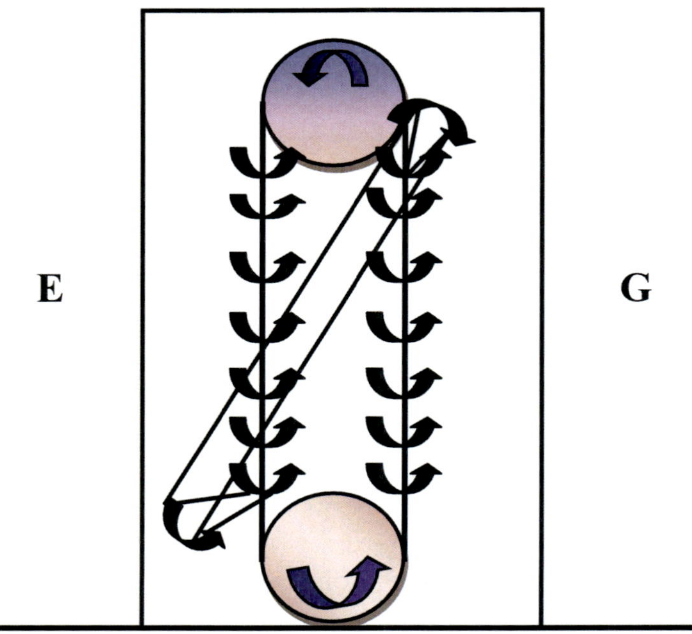

Figure 4.2 Buckets with atoms on a wheel experiment. The hypothetical system shows the effect of gravity on the absorption of light waves by atoms located at different heights. (Courtesy Sir Hermann Bond, KCB, FRS, UK, The theory of gravitation, Virat Surya- January 2005)

Einstein extended the results of the example to the time measured by atomic clocks. The consequence of his conclusion was that there is no absolute time. Time is always a value manufactured by clock. According to him, if clocks run differently at the top and bottom of the tower, then time must run differently at those locations. This implies that relationship between the local measured time and proper time is different at a different height. We agree with this fact that different clocks will show different times at different heights because the potential and kinetic energy associated with the clock measuring device is altered. However, measured time does not modify the elapsed or absolute time. Absolute time is the time measured when the distance between the source and the observer do not change with respect to time; this reasoning is similar to the true speed of light measurement as described in Section 2.2. Also, the source and the observer have the same potential and kinetic energy configuration. One of the side effects of Einstein's theory of gravitation is that just as measured time is different by clocks located at different heights, the measured length and mass of objects in motion is different. The measured length, mass, and time is related by the Lorentz transformation equations. We shall later explain that the Lorentz transformations are not required because these are virtual effects. In reality time, length and mass of objects are invariant.

In the next section, we will explain why concept of time dilation is incorrect and is detrimental to all natural phenomena. We will prove that time dimension is immutable and unmodified by any means. We stipulate this in spite of the observed differences in the measured time displayed by two clocks located in two different systems moving with respect to each other. We believe that the time dilation described by Einstein and others is an apparent or virtual effect. Absolute time is an invariant dimension because no physical processes, natural or artificial can modify wear-out mechanism of bio-degradable processes in a consistent and predictable fashion.

4.6 Time Dilation

Time dilation is a phenomenon related to the difference in clock time when it is measured between two separate frames of reference. The STR postulates investigated time dilatation by discussing time measurement using two clocks. One clock was located on the Earth's surface. An exactly identical measurement was assumed to be performed in an orbiting spacecraft surrounding the Earth at a speed of **25000 km/Hr**. The speed of a spacecraft is very high compared to any physical traveling object, but nowhere near the speed of light which is $\mathbf{3 \times 10^8}$ **m/s** or $\mathbf{2.998 \times 10^5}$ **km/s**. The clock consists of two mirrors separated by a distance of **1m**. A light source and the photo sensitive cell were fixed on one mirror facing the other. Both mirrors are located such that the direction of the light is orthogonal to the trajectory of the spacecraft. A simplified functional diagram of the mirror clock, and its associated principle, is illustrated in Figures 4.3a and 4.3b.

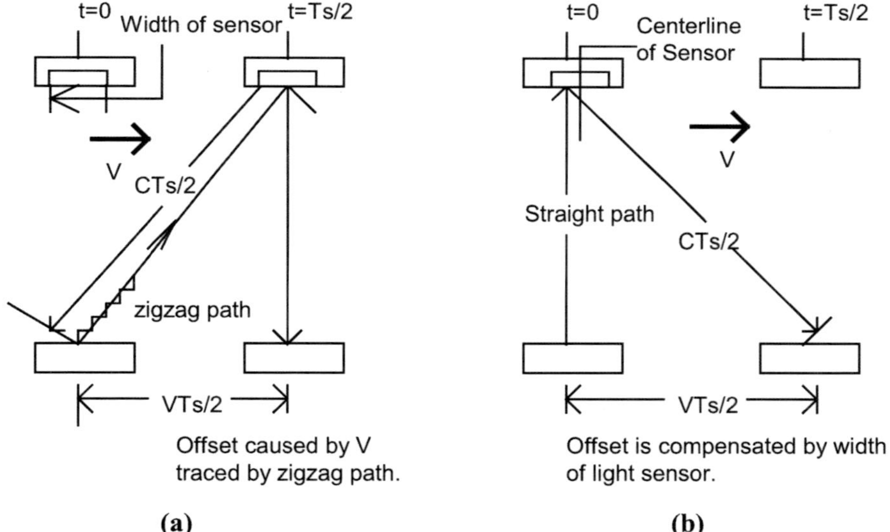

Figure 4.3 Path of light in a moving spacecraft. (a) The path of light according to Einstein's special theory of relativity. (b) The path of light according to the **Skylativity**® theory.

The clock time is measured by the formula:

$$\mathbf{E_t = 2L_s/c} \qquad \textbf{(4.7)}$$

where $\mathbf{L_s}$ is length of the stick in meters which is equal to **1m** and **c** is the speed of light

$= \mathbf{2\ m / 3 \times 10^8\ m/s = 0.67 \times 10^{-8}\ s}$

Because the speed of light **c** is assumed to be a constant, according to the explanations in several references, the time measured by a clock in an orbiting spacecraft will be slower than one on Earth as seen by an observer on Earth. We disagree that time dilatation should occur. The path of a ray of light (photon), in a zigzag pattern, as described in the relativity reference book by Roger Penrose [15] and by Einstein, is not correct. A flaw is that a photon does not have any mass. Therefore, the light wave will continue its journey in a direction normal to the spacecraft. It will hit the surface of the mirror at a point which is offset in direction lagging to the direction of the movement of the spacecraft. Figure 4.3a depicts the path of light similar to the path of light from a flash light in a moving vehicle described in a text book by Raymond Serway and John Jewett [22] and Figure 4.3b illustrates the modified paths. Also, our observation is that the offset should correspond to the distance of the time of travel of the light from mirror **A** to **B** multiplied by the speed of the spacecraft:

Offset $= \mathbf{0.67 \times 10^{-8}\ s \times 25000\ km/Hr = 46.52\ microns!}$

For the sake of discussion, we will continue the analysis stating that a virtual time dilatation occurs. The clock time measurements inadvertently indicate the slowing of the measured time. The rest of the analysis from reference [15] should remain the same and applies to this situation in the same way:

i. e. $(cT_s/2)^2 = (vT_s/2)^2 + L_s^2$ **(4.8)**

$T_s = (2L_s/c) / (1 - v^2/c^2)^{1/2}$

$T_s = T_e/(1 - v^2/c^2)^{1/2}$ **(4.9)**

where $\mathbf{T_e}$ is time measured by a clock on the Earth
and $\mathbf{T_s}$ is time measured by clock on spacecraft

Let us examine what should happen as the speed of the spacecraft increases. One effect is that the offset distance should increase. The clock will not function when the photo sensitive cell can no longer detect reflected beam due to higher offset. Offset distance greater than the width of the sensor. Suppose that one can ascertain that light is detected by repositioning the photo-sensors accordingly. Now, when the speed of the spacecraft becomes the same as the speed of light, the time dilation will be so high that $\mathbf{T_s}$ will be infinity and in no finite time will the light be able to reach the other mirror. The effect of the time dilatation can be diminished by decreasing the space between the mirrors. Ultimately, the distance between the mirrors should be reduced to zero to satisfy the equation (4.8). We like to name this effect the collapsing clock, due to the time dilatation. We think the experiment should be revisited in modern times when high technology equipment is available. This equipment can perform the measurements with very high accuracy.

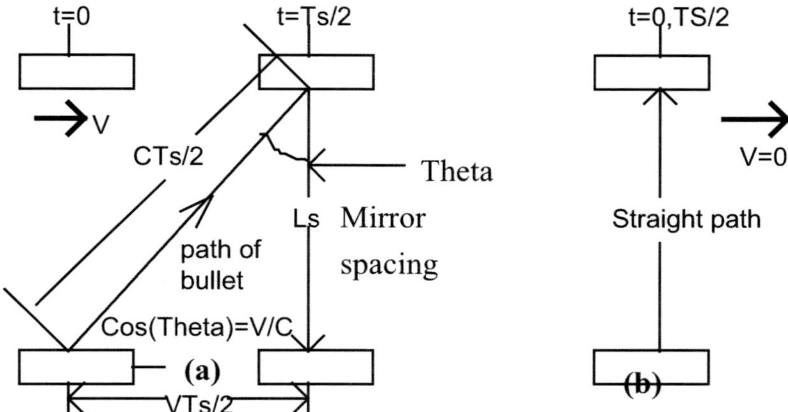

Figure 4.4 Path of a bullet fired in a spacecraft. (a) The bullet fired in a moving spacecraft. (b) The bullet fired on the Earth, ignoring the effect of the Earth's rotation on its axis and the orbital motion surrounding the sun.

We would like to recommend a slight modification to the clock. For this hypothetical clock, the light source should be replaced by a slower speed object, such as a bullet fired from a gun. Figure 4.4b illustrates the path of a bullet fired from a gun when the mirror is sitting on the Earth's surface. If the clock is located on the Earth and the bullet is fired exactly perpendicular to the Earth's surface, it should hit the mirror above it. The same experiment is performed in a spacecraft traveling at the speed of **V km/Hr**. The path of the bullet is illustrated in Figure 4.4a. Once again, the time dilatation should occur. At this time, the bullet should hit in a direction aligned with travel of the spacecraft. The path of the bullet should look inclined similar to the path of light described in the text book [15].

An undesirable effect of the time dilatation is that when the speed of any material object approaches or becomes comparable to the speed of light, the relative mass of the object increases toward infinity. According to Einstein's theory and the Lorentz transformation, it is established that the relative mass can be computed by the formula:

$$\mathbf{M@v = M@s/(1\text{-}V^2/c^2)^{1/2}} \qquad \textbf{(4.10)}$$

Meaning that in the limit:

$$\lim_{v \Rightarrow c} \mathbf{M@v - M@s \Rightarrow \infty}$$

where **M** implies the mass of the object
c implies the speed of light
V implies the high speed of the object
S implies the stationary condition

However, the time dilatation does not account for the case where the real mass is decreasing as the speed of the object is increasing. The effect of the dilatation can be alleviated by reducing the rest mass of the accelerating object. In fact, when the matter with the finite rest mass reaches the speed of light, it will begin emitting the light. In that process, a portion of the mass is consumed. Based on the conservation of energy, the relation between the initial mass of matter and the final mass can be expressed by the equation (4.10) and (4.11). As a result of the light emission, the kinetic energy of the initial mass is decreased. Therefore, the final mass $\mathbf{M_f}$ is less than the initial mass by quantity $\mathbf{M_v}$, which is consumed mass. The final mass attains a velocity $\mathbf{V_f}$ which should be less than **c**. For this discussion, we shall assume that the entire consumed mass is transformed into light wave energy. From the laws of the conservation of momentum and energy, we get:

$$\mathbf{M_i - M_f = M_v} \qquad \mathbf{(4.11)}$$

$$\mathbf{M_i \times c = M_f \times V_f + M_v \times c} \qquad \mathbf{(4.12)}$$

$$\mathbf{(1/2)\, M_i \times c^2 = (1/2)\, M_f \times V_f^2 + M_v \times c^2} \qquad \mathbf{(4.13)}$$

In practice, many events occur for which matter attains the speed of light. However, the life time at that speed is very short. Also, the speed can't be sustained for large distances. The main cause is that matter starts translating into light by its nature. Einstein's theory of relativity clearly did not describe this fact. A good example is that a nuclear reaction can be triggered by an atom with a relatively smaller mass and with a sufficiently high speed, in order to bombard a nucleus of a very heavy radioactive material, plutonium. In such reaction it is known and described by Einstein himself that massive quantity of light energy along with many other forms of energies is released. Also, it is well known that once the reaction is started, it leads to chain reaction and cannot be controlled. A good observation is that under such condition, the participating elements in the nuclear reaction attain an unstable state. Therefore, it is safe to say that whenever matter with a finite mass approaches the speed of light, it will be unstable.

One of the desirable effects of the time dilatation is that matter does not age if it travels at the speed of light. Many scientists claim that even the biological aging process of the living species, such as humans; will slow down if they travel at the speed of light [14]. Refer to the description of the ultimate space traveler in [2]. We disagree that matter and life will cease to age when they acquire the speed of light. Because of the refined path of light in the mirror clock, we claim that the time dilatation does not occur. Dr. Brian Greene tells in his book, Elegant Universe, that a photon never ages [14] because its speed is **c**. We would like to propose an experiment that proves that light waves or photon particles dissipate energy and can't sustain existence unless the wave energy is refurbished from its source. In our experiment a light bulb is located between two mirrors that are parallel and face each other, similar to the mirrors in a hair dressing saloon. In the night darkness, if we switch the power to the light bulb on, the bulb glows and several images of the bulb can be seen in the mirrors. However, if we switch the power to the bulb off, the room becomes dark immediately. If the photons emitted from the light do not age, the room should remain lit. Therefore, this experiment provides a proof that even photon particles, which travel at the speed of light, have a limited life span, and in that sense, the particles are aging. They lose their illuminating power.

In the next section, we shall consider a case of multiple images of quasars and distant galaxies as seen from a telescope on the Earth. It was suspected that this multiple images are observed because light arriving from those distant celestial objects suffered bending caused by gravitational force of nearby galaxies and stars. In reality we shall prove that it is not the case. The correct reason for bending is that light rays arriving from the objects were refracted by the Earth's atmosphere and the gas cloud surrounding the galaxies. At this point it is sufficient to state that several physicists have spend their life time to analyze multiple images of distant galaxies based on gravitational lensing effect has not produced very promising results as per our claims.

4.7 Gravitational Lensing

For several years it was believed that multiple images of remote galaxies were observed in telescopes because light detected from distant celestial objects from those galaxies was bending. The light was bent due to gravitational field of nearby galaxies. As explained in Section 4.6, in reality the multiple images are occurring as a result of the refraction of light when it passes through clouds of gases covering those distant galaxies. In astronomy to measure distances of stars stellar parallax technique is extensively employed. The success and accuracy of stellar parallax method is based on the premise that light does not bend by any force except reflection and refraction of light through ordinary glass lenses. Figure 4.4 illustrates the principle of stellar parallax technique. It is clear that measured distance of nearby stars by this method will be reduced if light arriving from the star is bent by gravitational field of nearby star. Further, it is evident that serious measurement error should occur if light will be bent other than its passage through lenses of measuring telescope [17]. It is evident from Figure 4.5 that if it is assumed that light is bent by the gravity of a nearby star; it will result in a reduction in the measured distance of the star. The star should appear closer than its actual distance.

As per our computation described in Section 1.2, the predominant bending effect is the refraction of light through the Earth's atmosphere and the refraction of light through gas clouds of intermediate galaxies. The main reason supporting our claim is light rays arriving from a distant star do not have a fixed center of gravity and lacks rest mass. If light rays bend by force of gravity, it should obey Newton's law of universal gravitation. The law states that every massive particle in the universe attracts every other massive particle with a force which is directly proportional to the product of their masses and inversely proportional to the square of the distance between them.

In symbolic notation

Every point mass attracts every single other point mass by a force pointing along the line intersecting both points. The force is directly proportional to the product of the two masses and inversely proportional to the square of the distance between the point masses:

$$F = G\frac{m_1 m_2}{r^2}\ , \qquad \textbf{(4.14)}$$

where

F is the magnitude of the gravitational force between the two point masses,
G is the gravitational constant,
$\mathbf{m_1}$ is the mass of the first point mass, $0 < m_1 < \infty$ (added by Shailesh Kadakia),
$\mathbf{m_2}$ is the mass of the second point mass, $0 < m_2 < \infty$ (added by Shailesh Kadakia), and
r is the distance between the two point masses.

It is evident from above equation that it does not specify the value of force for which any object have zero static mass. We suggest that the equation should restrict the mass values to greater than zero and less than infinity. For mass value of zero or infinite the distance between the two point masses r is not determined. Further, the distance r between two point masses, he implied the distance between the centers of gravities the two point masses.

The center of gravity of light waves is neither static, nor even quasi-static because source of light the star is neither in static frame of reference nor quasi-static (dynamically static) frame of reference. The positions of light rays arriving from stars that are measured are continuously variable. Further, the bending of rays could be affected by gravitational field of more than one star. In that case the method will never be able to provide measurements with minimum required accuracy. You will soon discover that intuitively our argument is very convincing. The primary benefit of principle that light does not bend by the force of gravity is one need not apply correction factors of complicated gravitational bending effects on light waves to locate position of stars by stellar parallax method. One obscure fact about gravitational lensing is scientists are unable to explain why not all the objects exhibit multiple images in the universe when viewed by telescope. If gravity is a cause for the bending of light, all of the objects should portray multiple images. It is not obvious that only some of the quasars of the distant universe exhibit multiple images. On the other hand, if we believe that the refraction of light rays, as they pass through the atmosphere and gas clouds, are the reason for multiple images, then it makes sense that not all celestial objects will have multiple images. The angle of the incidence of rays from many celestial objects may result in the total internal reflection based on the index of the refraction of the atmosphere. Therefore, only a few of the many celestial objects would show multiple images.

One of the major advantages of the principle laid out by us that light coming from distant star does not bend by force of gravity of other stars is it allows to map position of stars in sky without worrying about continuously applying correction factor for observed displacement caused by the disturbing force of gravity. The value of our inventions should not be under stated because it makes dramatic difference in ability of technology and software programs to predict future positions of stars based on past positions. Also, the light arriving from some of the galaxies with dense gas cloud atmosphere will be passing through and suffer from refraction, and show multiple images.

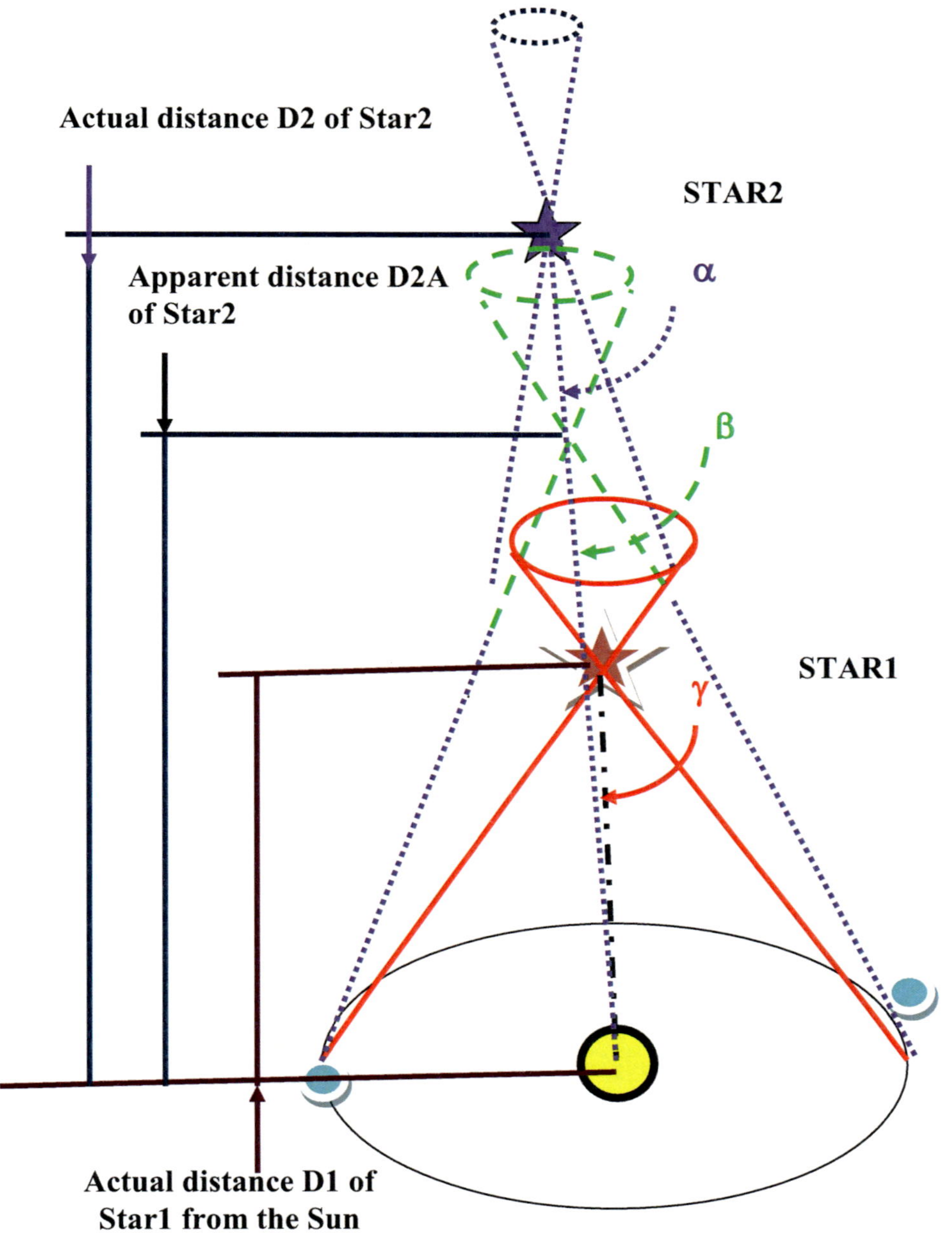

Figure 4.4 Effect of Gravitational Lensing: stellar parallax technique. The apparent position of a distant star is affected by the force of gravity from a nearby star, as per Einstein's general theory of relativity.

During our visit to India in fall of 2009, we had opportunity to meet India's prominent physicist and Emeritus Professor Dr. Jayant V. Narlikar from Inter University Centre for Astronomy and Astrophysics (IUCAA) in Pune, India. While our conversation of **Skylativity**®, we had detailed discussion on the bending of light in the presence of intense gravitational field of dense celestial objects. When we presented the viewpoint that light rays with zero rest mass does not bend by force of gravity, because they do not have center of gravity. He responded that in the absence of a strong gravitational field, the light rays should follow null-geodesic trajectory, meaning they travel in a straight line. He mentioned that gravity warps space and time in proximity of intense gravitational fields and, therefore, the scientists from the modern world believe that light rays obeys the rules of non-Euclidian geometry and may bend, or in other words, the trajectory of light rays may be curved. From this book's point of view, space without matter should not be warped because force of gravity should not distort any geometry without shape and mass. We ended our conversation for the fact pending further investigation.

4.8 Summary

This chapter gave a good overview of limitations of Einstein's theory of relativity. It explained some of the failures of the postulates of the STR, by the detailed analysis of the physics behind the operation of the common incandescent bulb lamp. The close analysis of the process of light emission from a common incandescent light bulb, invented by Edison, revealed the fact that it contradicts the postulates of Einstein's STR, that infinite energy is required to accelerate the particles at rest to the speed of light. In Section 4.2, we showed that mass of tungsten filament does not change even after radiation of light waves for its operational life. In Section 4.3, we discussed why the Lorentz transformation may result in the mass paradox for the matter resident in different inertial systems. In Section 4.4, we introduced concept of mass as being a vector quantity which has two components a real rest mass and an imaginary mass due to kinetic energy of the object. Obviously real mass directly affects location of the center of mass (center of gravity) for the object. The imaginary mass or energy mass components help to locate future position of the object. In Section 4.5, we discussed the flaws of theory of gravitation from Einstein.

In Section 4.6, it was explained why the constancy of the speed of light and the time dilation were incorrect concepts. We made this conclusion based on the analysis of examples of paths of light in moving and stationary spacecraft. Further, we proved that the multiple images of quasars and distant galaxies are observed by the refraction of light rays, as they pass through the Earth's atmosphere and the gas clouds surrounding the galaxies, rather than the bending caused by the gravitational field developed in the vicinity of the galaxies. In Section 4.7, we explained using an example stellar parallax technique to prove that the bending of light by the force of gravity will produce devastating measurement errors in distant computation of nearby stars. In the next chapter, we shall examine limitations of the Lorentz transformation to compute time, length and mass measurements in inertial systems. Also, we shall explore the boundary conditions and modifications required for computation of time, length and mass due to variations in the speed of light among different frames of reference.

5.

Simplified Lorentz's Transformations

In this chapter, our focus is to discuss the complexities involved with the Lorentz transformation and to explain why it does not make sense to use those transformations for computing the scales of mass, length, and time among the different inertial systems **S** stationary and **S'** moving with velocity **v** with respect to **S**. In Section 5.1, we shall understand the reasoning behind the development of the Lorentz transformation equations to compute the different scales of mass, length, and time among two systems **S** and **S'**. We shall see that the Lorentz transformation was derived on the premise that the speed of light remains constant when measured in both systems **S** and **S'**. If this assumption is invalidated, the equation does not provide the computation of the correctly scaled quantities of mass, length, and time. We shall begin our discussion by describing time/space coordinate translation diagrams for the translation of coordinates of the length between the two systems **S** and **S'**, as illustrated in Figure 5.1a, 5.1b, and 5.1c.

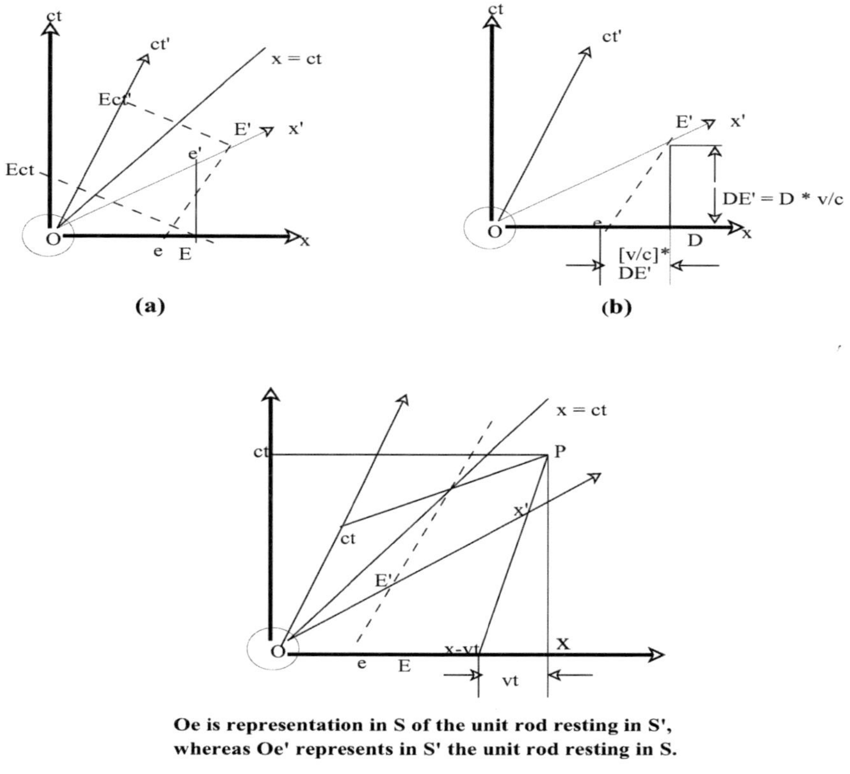

Figure 5.1 The speed of light constant c, Lorentz transformation of a world point P. (a) The units of space and time in **S (E, Ect)** and in **S' (E', E'ct')**. (b) The calculation of the ratio **E'/e**. (c) The Lorentz transformation of coordinates of a world point **P**.

In Section 5.2, we shall discuss the scenario were **S'** is approaching **S** with velocity **v km/s**. For the purpose of analysis, we shall show the effect of variable **c** on time and space coordinates, using the translation diagrams 5.2a, 5.2b, and 5.2c for the approaching scenario. We shall discuss the boundary conditions for the application of the modified Lorentz transformation with the correction for the variation in the speed of light on the computation of the units of length.

In Section 5.3, we shall discuss time and space coordinate translation for scenario where system **S'** is receding away from system **S** with velocity **v km/s**. For analysis of this situation, we shall draw the time/space coordinate translation diagrams 5.3a, 5.3b, and 5.3c. We shall discuss the boundary conditions for the application of the modified Lorentz transformation with the correction for the variation in the speed of light on computations of the units of length. In Section 5.4, we shall summarize the discussion on the modified Lorentz transformation on the measured scales of length, mass, and time among the different inertial systems for the approaching and receding scenarios.

5.1 Lorentz Transformation and Virtual Time Dilatation

The Lorentz transformation derives the relations between lengths and times in various inertial systems. These transformations are based on the principle that the speed of light has the same value when measured with rods and clocks of the same kind, in all inertial systems [7]. Einstein utilized these transformations to determine the contraction of the length of the rod and the dilatation of the measured time for two inertial systems. The system **S'** moving with velocity **v**, while system **S** is at rest.

The expressions for length are:

$$\mathbf{l = l_0 \,(1\text{-}v^2/c^2)^{1/2}} \qquad \textbf{(5.1)}$$

equation (74) on page 248 [7].

This states that the measured length of rod is shorter in the moving system **S'** than the measured length in system **S**, at rest. Conversely, a road at rest in **S'** appears contracted by the same amount when measured in **S** moving at the speed –**v**.

The expression for time is:

$$\mathbf{T = T_0 / \,(1\text{-}v^2/c^2)^{1/2}} \qquad \textbf{(5.2)}$$

equation (75) on page 250 [7].

This states that time interval in the system **S'** appears lengthened if measured in **S**. More generally speaking, viewed from any one system, the clocks of every other system moving with respect to former appear to be losing time. We agree that time appears to be slow. The important key word is appear that means it is a virtual effect and not real. The time dilatation is reciprocal to the contraction in length. For the sake of discussion, we will investigate dilatation and length contraction among inertial systems with an assumption that the speed of light is not constant and changes according to the Galileo transformation.

Our objective is to derive similar expressions for length and time transformation with an exception that velocity of light is different in two systems. The difference is related by the Galileo transformation:

$$c' = c - v \qquad \textbf{(5.3)}$$

equation (42) page 125 [7]

When Lorentz derived the relations, he considered two situations. He measured the length of the rod in system **S'** moving with velocity **v km/s** with respect system **S**. In his discussion system, **S** was at rest and considered to be the reference system. He concluded in his analysis that the relations differ only in the sign of **v** when he switched to **S'** as the reference system. In either scenario, the measured length was shorter and the measured time was longer than those quantities for the reference system. Two distinct situations should be considered. In the first case, the velocity difference is **v km/s** and its direction is approaching the other system. In the second situation, the velocity difference is **v km/s** and its direction is receding away from the other system. These approaching and receding scenarios are more important in space because in the universe the changes in the sign **v**, due to the different reference system, is of lesser relevance than the concept of the proximity and the spatial coarseness of the heavenly objects. In the next section, we shall consider the world line diagrams for the rod of unit length for the instance of the approaching systems **S'** moving with velocity **v km/s** toward system **S**.

5.2 Approaching System

The world lines of a rod of unit length in two different inertial systems **S** and **S'**, based on the constancy of the speed of light (Figure 114 (a), (b), (c) page 234 [7]), are reproduced in Figures, 5.1a, 5.1b and 5.1c. Also, Figures 5.2a, 5.2b and 5.2c illustrate the world lines of a rod of unit length in two different inertial systems **S** and **S'**, based on special **Skylativity®** theory and the approaching scenario. In these figures, **S'** is approaching **S** at the velocity **v km/s**. The figure has a difference for the time interval in **S'** which is represented by **(c-v)* t'**, instead of constant value **ct'** as in Figure 5.1a. In Figure 5.1a, the world lines for light in both systems **S** and **S'** are represented by equations $x = ct$ and $x = ct'$ coincident lines. In Figure 5.2, the world lines for light in two systems, **S** and **S'**, are different and represented by equations $x = ct$ and $x' = (c-v) \times t'$ respectively.

To simplify the procedure, the conventions and nomenclature between Figure5.1 and Figure 5.2 are kept the same. You may refer to the details of the derivation in [7].

Instead of repeating the geometrical discussion, we will focus on the difference in the formulae and discuss the effects on the results due to modified principles.

Let **E** represent a rod of unit length at rest in system **S**:

E' represents the unit of length in the system **S'**
E represents the intersection of world lines of end **E'** to x-axis in system **S**
e' represents the intersection of world lines of end **E** to x'-axis in system **S'**

As indicated in Figure 5.2b, the effect of the different speed of light in **S'** is reflected in the computations of side **DE'** and side **eD**. The multiplying factor in Figure 5.1 (b) **v/c** is replaced by **v/(c-v)** in Figure 5.2 (b). The computed values for the sides are **D × [v/(c-v)]** and **DE' × [v/(c-v)]** in Figure 5.2 (b) which are different than **D × (v/c)** and **DE' × (v/c)** from Figure 5.1 (b). Conventionally, ratio **v/c** is represented by the Greek letter **β**. Let:

$$\gamma_a = v/(c-v) = (v/c)/(1-v/c) = \beta/1-\beta \qquad (5.4)$$

where γ_a is a parameter similar to β

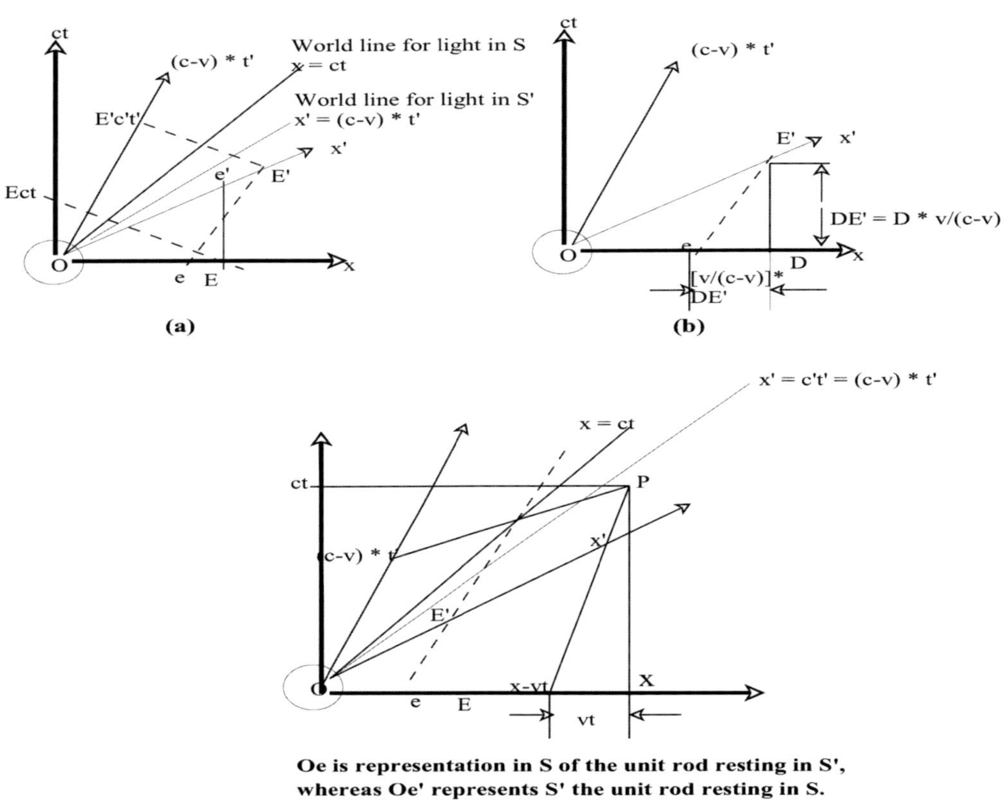

Figure 5.2 VSL, approaching scenario: Lorentz transformation of world point P. (a) The units of space and time in **S (E, Ect)** and in **S' (E', E'c't')**. **S'** moves at the speed **v** with respect to **S** which is the system of reference. (b) The calculation of the ratio **E'/e**. (c) The Lorentz transformation of coordinates.

There are two different types, γ_a for the approaching scenario and γ_r for the receding scenario. It will be evident from the detailed analysis, that the equations were derived based on the case for the non-constancy of the speed of light (the new postulates) and is identical to the original transformations, if the factor in terms involving **v/c** or β are replaced by the corresponding factor **v/c-v** or γ_a for the approaching system. This is good news because the upper limit on the differences between the two computations is no worse than the relationship between functions β and $\beta/1-\beta$. For the small velocity difference **v** between the two systems, compared to the speed of light **c**, then we arrive at the Galileo transformation. The interesting results are seen for the approaching system and the non-constancy of the speed of light assumption. For a new transformation γ_a increases more rapidly than β. When β is ½ when **v** = **c/2** the speed of light $\gamma_a = \infty$. This says that, for approaching systems, one cannot go beyond half the speed of light because the time dilatation is infinite. Also, the length of the rod will be reduced by 100% and the rod will disappear or cease to exist. Therefore, we would say Einstein's computations were optimistic and require a lot of corrections. In the next section, we shall discuss world-lines for receding scenario in which system **S'** is moving away from **S** at a speed of **v km/s**.

5.3 Receding System

The world lines of a rod of unit length in two different inertial systems, **S** and **S'**, based on constancy of the speed of light (Figure 114(a), (b), (c) page 234 [7]), are reproduced in Figures, 5.1a, 5.1b and 5.1c. Also, in Figures 5.3a, 5.3b and 5.3c, the world lines of a rod of unit length are illustrated in two different inertial systems, **S** and **S'**, based on the improved postulate for the special **Skylativity**® theory and the receding scenario. In these figures, S' is receding away from **S** at velocity **v km/s**. Our figure has a difference for the time interval in **S'** which is represented by **(c+v)* t'** instead of constant value **ct'**, as in Figure 5.1a. In Figure 5.1a, the world lines for light in both systems, **S** and **S'**, are represented by equations **x = ct** and **x = ct'** coincident lines. In Figure 5.3, the world lines for light in two systems, **S** and **S'**, are different and represented by equations **x = ct** and **x' = (c+v) × t'** respectively.

To simplify the procedure, the conventions and nomenclature between Figure5.1 and Figure 5.3 are kept the same. You may refer the details of the derivation in [7].

Similar to the approaching scenario, instead of repeating the geometrical discussion, we will focus on the difference in the formulae and discuss the effects on the results due to modified principles.

Let **E** represent a rod of unit length at rest in system **S**:

> **E'** represents the unit of length in the system **S'**
> **e** represents the intersection of world lines of end **E'** to x-axis in system **S**

e' represents the intersection of world lines of end **E** to x'-axis in system **S'**

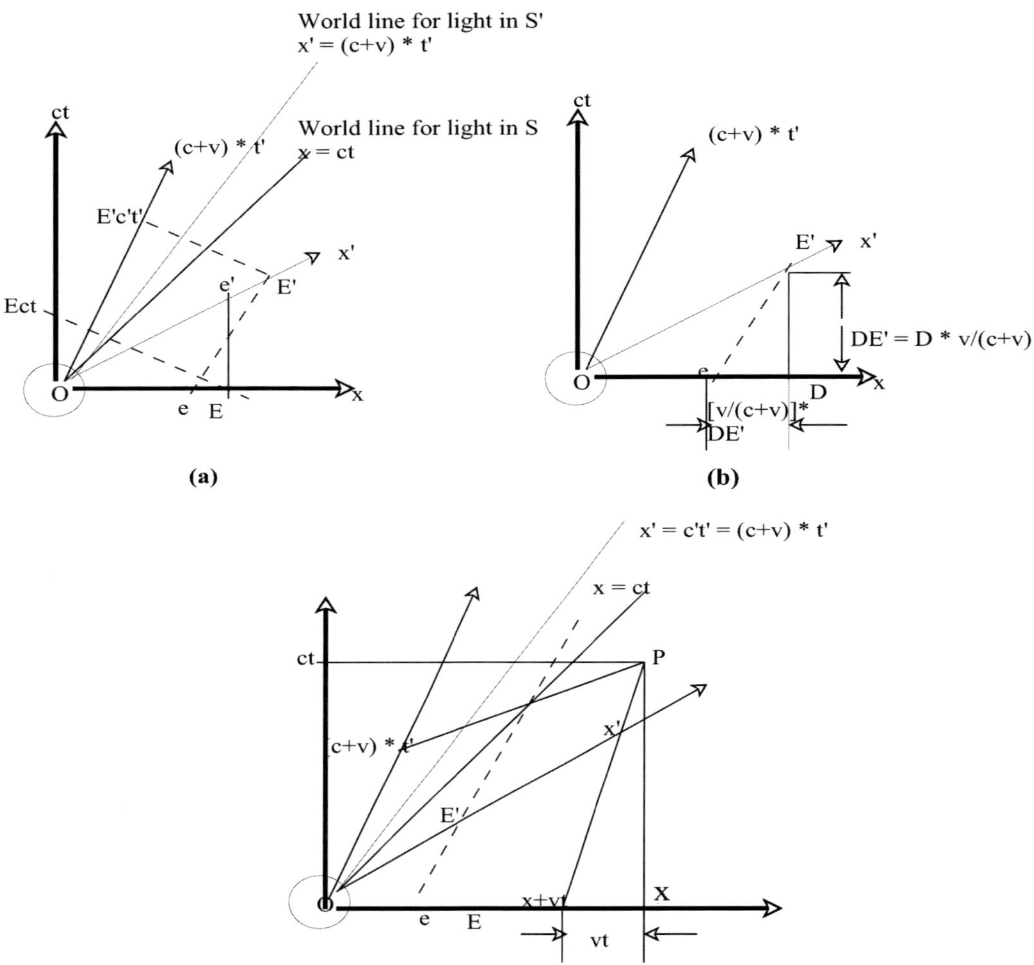

Oe is representation in S of the unit rod resting in S', whereas Oe' represents in S' the unit rod resting in S.

(c)

Figure 5.3 VSL, receding scenario: Lorentz transformation of world point P. (a) The units of space and time in **S (E, Ect)** and in S' **(E', E'c't')**. **S'** moves at the speed **v** with respect to **S** which is the system of reference. (b) The calculation of the ratio **E'/e**. (c) The Lorentz transformation of coordinates.

As indicated in Figure 5.3b, the effect of the different speed of light in **S'** is reflected in the values for the sides, **D × [v/(c+v)]** and **DE' × [v/(c+v)]** instead of **D × (v/c)** and **DE' × v/c)**, as in Figure 5.1b. Conventionally, the ratio **v/c** is represented by the Greek letter β:

Let $\gamma_r = v/(c+v) = (v/c)/(1+v/c) = \beta/1+\beta$ **(5.5)**

where γ_r is parameter similar to γ_a

There are two different types, γ_a for approaching scenario and γ_r for receding scenario. It will be evident from the detailed analysis that the equations derived based on non-constancy of the speed of light (the new postulates) resemble and is identical to the original transformations if the factor in terms involving **v/c** or $\boldsymbol{\beta}$ are replaced by corresponding factor **v/c+v** or γ_r for receding systems. This is good news because the upper limit on the differences between the two computations is no worse than the relationship between functions $\boldsymbol{\beta}$ and $\boldsymbol{\beta/1+\beta}$. For small velocity difference **v** between the two systems compared to the speed of light **c**, we then arrive directly at the Galileo transformation. Notice that for receding system γ_r is always smaller than value for $\boldsymbol{\beta}$. This implies that the reduction in length is smaller for the case of the non-constant speed of light than the reduction for the case of the constant speed of light. Also, the time dilatation is always higher for receding system than for the case of the constant speed of light.

One of the direct significances of the Lorentz transformation for time dilatation is the effect on life time computations of very high energy particles. To study the decaying characteristics of particles with the speed nearing the speed of light speed, scientists measure the time of decay. Usually, their life time is very short in static frame of reference. But its internal clock slows down considerably due to dilatation when measured in its own frame of reference. The results of life time computation could be drastically different using the new transformation. Also, the results vary depending on the situation where the particles life time is measured when they are traveling toward the measuring device or away from it. The main advantage of measuring life time in its frame of reference is it allows analysis of data spread over a larger time interval than a static frame of reference. We would like to emphasize the fact that the concept of time defined by us is local to our solar system. It has no meaning in the absolute sense. Therefore, relativistic life time measurements in a different frame of reference may differ. However, the parameter of interest should be the decay rate based on the energy content of the particle. The decay rate should be independent of the life time measurement in a frame of reference. We are implying that the quantity of matter remaining after decay of certain number of time constants should be the same regardless of frame of reference.

5.4 Summary

In this chapter we began with discussion of word lines for a rod of unit length as it relates to two systems **S** and **S'**. It is stipulated for the purpose of discussion that system **S** is stationary and system **S'** is moving with the speed **v km/s** either toward **S** or moving away from **S**. In Section 5.1, it was pointed out that the Lorentz transformations, derived on the basis of the constancy of the speed of light, are not adequate. These transformations are utilized to compute the length, mass, and time for two different inertial systems **S** and **S'** moving at the speed **v** relative to each other. To simplify details of discussion we skipped geometrical details and instead focused on what happens if we made an assumption that the light speed does not remain constant but follows the Galileo transformation. We explained that we consider two separate scenarios approaching systems and receding systems. In the universe, the cases for approaching and receding systems play a more important role than systems with positive or negative speeds.

In Section 5.2, we considered the world line diagrams for the rod of unit length where system **S'** is approaching **S** at the speed of **v km/s**. Our analysis suggests that when the speed of light is assumed to be variable for the approaching scenario, time dilation approaches infinity and the length of rod is reduced to zero length when the velocity of **S' v** exceeds **c/2 km/s**. This proves that the Lorentz transformation computes unrealistic values for measured length and time if his equations are extended for the instance of light speed variations consistent with the Galileo translation, and **S'** is approaching **S**. In Section 5.3, we derived the expression for the measured length and time for the rod of unit length when **S'** is receding from **S** with the speed **v km/s**. We concluded that for this case reduction in length is smaller than the corresponding reduction for the case of the speed of light constant. Also, the time dilation for the case with the variable light speed is smaller than the time dilation value when the speed of light is assumed to be constant.

In fact, the original equations for transformation provide optimistic figures for both measured time and measured length. Revised expressions for transformations are derived in Sections 5.2 and 5.3 based on refined postulate velocity of light **c** and **c'** is different in two systems **S** and **S'** moving with velocity **v**. The new transformations are derived for two scenarios, when inertial systems are approaching toward and receding from, each other. The transformation predicts that measured time slows down considerably and measured length is rapidly diminishing as the speed difference between two systems approaches half the speed of light. Since the Lorentz transformation equations does not provide realistic values and in view of the fact that time dilation is a virtual effect, we conclude that the Lorentz transformation equations should not be employed in its present form to compute length contraction and time dilation. Next, we briefly discussed the time dilation as it relates to the life time and decay characteristics of the high speed and high energy particles, such as neutrinos. We concluded that it is not relevant that the larger spread in life time, experienced from their frame of reference, provides more meaningful results than monitoring the energy content of the particle. We implied that the decay rate and the quantity of remaining matter do not depend on measurement in different frames of reference.

In the next chapter we shall discuss applications of **Skylativity®** theory to space age. We shall explain why it is not essential to build more advanced super colliders and particle accelerators that involve huge investments. We shall discuss that the funds utilized in construction of huge super colliders should be allocated to perform research in different areas of space activity. After that we shall show that Universal scale for mass, length and time exist and these measurements are independent of frame of reference.

6.

Applications of Skylativity® to the Space Age

In this chapter we shall describe various benefits of the new postulates of **Skylativity**® to the space industry for variety of applications.

In Section 6.1, we shall explain why it is not essential to construct a larger super collider than previous generation as a direct consequence of the postulates of K-theory. The dollars saved by not spending on the construction of larger super colliders, can be applied to fund projects that develop spacecrafts with much higher speeds. Later it will be evident that our universe is comprised of very coarse and dilute system of celestial objects stars and galaxies with huge scattered distances. Therefore, it is imperative to invent rocket engines that boost the velocity of travel for space vehicles, at much higher levels than the current speed of 11-15 miles per second. In Section 6.2, we shall show that as a result of the **Skylativity**® theory it may be possible to formulate Universal Unified Field Theory (UUFT) that would close all the gaps left behind by Einstein's theory of relativity. We could achieve this objective because the measures and scales of mass, length and time among different inertial systems have the same values. It does not require the Lorentz complex transformation equations to convert length, mass, and time values from one system to the other.

In Section 6.3, we shall explain an important phenomena life time (decay rate) of fast moving neutrino particles arriving from the sun. After reading the detailed explanation in this section, you will be convinced that neutrino survives not because of apparent increase in life time of these particles as per explained by proponents of Lorentz and others. The main reason neutrino particles survive because they are highly inert and stable. In the Section 6.4, we shall describe application of quarks in producing future weapon systems.

In Section 6.5, we shall touch on the future space expeditions and the suggested priority for NASA to plan for future projects. We will suggest that many experiments those are performed from space station may be performed from a human habitat on dark side of moon. Also, the universe can be observed without the interference of atmosphere similar to the Earth's atmosphere as the satellite moon does not have atmosphere. The mapping of the sky/universe will be much more accurate from the moon than from an observatory on the Earth. Further, to improve the safety of astronauts, we suggest that escape vehicles that will protect astronauts and allow emergency escape from spacecrafts in the event of accidents, should be designed.

In Section 6.6, we shall discuss the considerations involved in the design of spacecrafts with very high speeds to fractions of the speed of light. The primary need for such speeds arises from the fact that the distances of stars in our neighborhood are very large. The nearest star alpha century is 4.5 light years away! At present, at the speed of our spacecrafts, 11-15 miles per second, the travel mission takes six months to get to the nearest planet, such as Mars, and takes several years for a probe to reach Pluto. With spacecrafts with improved speeds, the travel time for these missions can be significantly reduced.

6.1 Reallocation of particle collider Resources

For many years, scientist tried to accelerate matter particles to a speed in excess of the speed of light. The organizations such as CERN (Center for Electronics Research in Nuclear physics) in Switzerland and SLAC (Stanford Linear Accelerator Center) in California constructed huge particle collider. The circumference of the Large Hadron Collider at CERN is 16.6 miles. The linear collider at SLAC is approximately two miles long. In a super collider, a very high speed and heavy particle (huge momentum) neutron or proton, is collided with a relatively low mass, high speed electron. The idea is to accelerate the electron further beyond the speed of light. In practice, the electron at certain high speeds, collides the heavy particle and smashes it into quarks. Therefore, transfer of the momentum to increase the speed of the electron is prohibited. At the same time, the light and other radiation waves are created that decrease the kinetic energy of the particles. The key concept of the new theory, that whenever any composite particle comprised of molecules and atoms is accelerated to the speed of light, it will emit light waves. This has a great financial advantage because the need for constructing larger than the previous generation super collider is diminished. Thus, multimillion dollars funding invested in vast construction of super collider projects will be saved. As stated earlier, as soon as a particulate of matter attains the speed of light with any means, it will emit light. In that process some energy is consumed. That will result in reduction of the kinetic energy and subsequently the speed of the particle.

It is interesting that one such project of constructing underground super collider at site near Dallas was abandoned in recent years. The project was cancelled because of the budget, as well as physical limitation of underground construction. Based on the findings from this book, it is obvious the construction project would not have served much purpose, which is a matter of pure coincidence. If the ideas presented in this book were published early by seven years the project would not have been initiated and would have saved the funds applied toward the construction to the greater extent. Nevertheless, the funds that would have been applied to build a huge super collider in future will be available for promoting other research, such as the development of rocket engines, beyond the speed of the existing generation of rocket engines.

6.2 Universal Time, Length and Mass Scales

Another significant advantage of the new theory is that the speed of light is not independent of frame of reference. Therefore, all the measurements of length and time hold well for different frames of references. Thus, the need for performing length and time measurements in a different frame of reference is eliminated. Also, it is not necessary to apply the Lorentz transformation equations for length contraction and time dilation computations among different inertial systems. This is an extremely valuable result because it simplifies a lot of computations. Further, in practice it does not make sense that one can accelerate one electron or one proton above the speed of light. The matter is comprised of molecules and atoms. There is no element, for which the atomic configuration is one electron or proton that exists in nature except the ions of hydrogen. Such ions do not have a stable state. Whenever an atom with more than one electron, is accelerated to the speed of light at the precise kinetic energy thresholds, the electrons in the outer shell orbits are excited. The excitation of the electrons causes the electron transition event and the radiation energy is released. As a result of the radiation, the kinetic energy or the speed of the charged particles is diminished. Therefore, to devise a particle accelerator that speeds up a single proton or neutron, above the speed of light is not profitable from a financial point of view. We do not discard the prior investments in particle accelerators. Several other particles and quarks were invented by studying the radiation spectra when high energy particles were collided.

6.3 Life Time of Neutrino

One of the direct applications of the **Skylativity®** theory is that one need not apply time dilation for decay phenomenon of fast moving neutrino particles. In April 2005, we attended a meeting at Boulder Astronomical Space Society (BASS). The topic of the meeting was Cassini Mission and landing of probe on Titan a Satellite of Saturn. The speaker, Mike Hotke from Ball Aerospace & Technologies Corp., presented an excellent review and shared many photographs from the two missions. The first ones, from Jet Propulsion Laboratory (JPL) of the satellite taken by the probe that was landed recently and the other photographs were previously taken by Hubble Space Telescope (HST). Hotke raised an interesting question about the life time of the high speed neutrino particles arriving from deep space, showering the Earth's atmosphere, and disappearing after a short distance of travel. He specified that NASA is utilizing the decay rate of these particles to determine the origin of the solar system and celestial objects. He said that when the high speed particles, with speed nearing the speed of light, have a very slow rate of decay in space because it is experiencing a very slow clock. He asked us to explain life time of neutrino particles based on **Skylativity®** theory.

We provided an answer to his questions. When the particles travel inside the atmosphere, their decay rate suddenly increases because the speed diminishes very rapidly, encountering friction and collision with gas atoms. We believe that the decay rate of the neutrino particles in space is very small because the decay time constant is very large. In fact, it has been discovered that the solar neutrinos have been detected with the expected flux and do not decay over the $\mathbf{500 \times c}$ distance to the Earth [27]. In this paper several models for decay rate or life times of neutrinos arriving from different sources and limit on decays are discussed. It turns out that in space there is no resistance to the motion of the particle. Therefore, the slow decay rate is not because the internal clock for aging the particle is slow, but it is that the time constant of the system is very long. If the clock were slow, then the particles should be able to travel a lot longer distance in the atmosphere. It is noted, that there is not any significant reduction in the speed of these particles, even after they enter the Earth's atmosphere. Usually, these particles have been found to disappear after the short distance travelled in the atmosphere. The main reason particles disappear after the short distance of travel, is they collide with gas molecules before they can make it to the surface of the Earth. Also, the majority of the neutrinos pass through the Earth without being detected because they do not interact with any matter.

6.4 Quarks: Source of Energy for Future Weapons

The **Skylativity**® theory could possibly have a significant impact on the design of the next generation of weapon systems. According to our explanation in Section 2.7, on revised energy computation in nuclear reaction allows calculating components of various forms of energy released during the reaction with high accuracy. If a reaction is invented that transforms protons and neutrons into quarks by means other than large Hadron collider a new class of weapons can be developed. The weapons will be based on a process of smashing protons and neutrons into fundamental particles such as quarks [5]. There is a potential for vast quantities of release of binding energy when neutron or proton are smashed into quarks in a weapon system fabricated as such and detonated. Further detail on the subject matter is skipped in view of the fact that it is outside the scope of this book. We wish to point out a fact that bombing of cities in Japan during Second World War caused loss of hundreds of thousands of lives. The vast loss is not due to the incredible amount of energy release but was attributed to the dangerous radiation effects those are very harmful to human tissues and life. A recent accident in nuclear plant in Chernobyl, Chechnya had caused loss of many lives for the similar cause. Hence, nuclear plants and factories of nuclear weapons are constructed in shielded facilities and extreme caution is observed in their operations.

6.5 Future Space Expeditions

Yet another application of the **Skylativity**® theory facilitates the design of a space vehicle at the speed of light. Einstein's conclusion that the relativistic mass of matter increases as the speed of mass increases is not correct. It is found that as the speed of the mass increases, the object gains kinetic energy. At the same time, the position of the object, in relation to other objects changes, results in a change of the potential energy. Essentially object trades kinetic energy to potential energy. The mass of object should not change. Previously, aerospace engineers were prohibited from designing very high speed aircrafts because the thrust required to accelerate matter in the proximity of the speed of light, increases exponentially. On the basis of this new theory, as well as the present state of technology, we believe that it is possible to design rockets with speeds ten times larger than the current speed. The primary limitation is how to decrease the speed of those space vehicles, when the spacecraft approaches the destination planet or satellite. The higher the speed of the space vehicle, the size and thrust required for retarding the speed from control rockets, will be high. This implies that the payload of the control rocket on the spacecraft will be high at the beginning of the journey. The payload requirement on the design of the spacecraft and retardation rockets is a loop back limit situation. Clearly new and improved theory identifies a need for redesign of spacecrafts.

At this point we would like to shift our focus to a slightly different topic, space program and proposed objectives of NASA in future. At present new director of NASA Dr. Michael Griffin has stated that he would like to send man on the moon to put back the space agency programs on space forefront after the Challenger's disaster. The success of the current project of the International Space Station and its ahead of schedule assembly should boost the confidence of NASA in Joint effort space programs. Therefore, NASA should consider a mission to construct a space station on the bright side of the moon. We shall explain why it should be on the bright side and not the dark side. Sending humans only to the moon will not be the appropriate objective, because that goal was already accomplished in the prior century.

We suggest that in this new millennium NASA's goal should be to use resources from coalition of Europe and Asia for success of new project similar to success of Space Station orbiting now. NASA should build a greenhouse on moon based on expertise and experience of International Space Station. Construction of a station on moon similar to greenhouse in deserts has many advantages. The risk on surface of moon is less than one which is orbiting in free space. This is true because the moon has reduced gravity environment as opposed zero on the space station. The gravity in the current International Space station is simulated gravity. Over a period of time astronauts can get used to low gravity. Gravity simulation will not be required for station on the moon. Also, landing of shuttle and/or capsule on the moon faces less stringent conditions. The moon has a lower gravitational pull than the Earth and no atmosphere. Thus, there is less impact thrust upon landing, and the surface temperature of the landing space vehicle should not rise. The plant and trees inside greenhouse laboratory on moon could provide transformation from Carbon dioxide into oxygen if the laboratory is located on the bright side of the moon. We believe that the closest non-Earth habitation for humans is highly feasible on our own satellite, rather than any other planet in the solar system or outside the solar system. The primary reason, it is difficult to overcome the speed and payload conflict for the shuttle design, because the high speed requires deceleration rockets with a lot of fuel to slow the speed of the spacecraft when the space vehicle approaches the destination planet.

If another laboratory will be constructed on dark side of the moon it will facilitate study of our universe and celestial objects with more ease, and allow performing measurements with high accuracy. The study of the celestial objects from the moon will be more precise than from the Earth because the moon does not have atmosphere. Therefore, locations of stars are much more accurate. An astronomical laboratory on moon will be highly useful and provide accurate measurements. Many astronomers believe that the refraction effects due to the Earth's atmosphere can be corrected in software when observing universal objects from the

Earth. This is not accurate and it is a complicated procedure. The primary reason is that the correction factor is not constant. Also, the different stars project different angles of incidence to the telescope and the stars are not stationary. Therefore, correction is an NP incomplete problem means real time correction is not feasible. Also, information arriving from different stars is uncorrelated in time because the instantaneous distance and the speed of each star are different and not related. Our proposal is that the light arriving from the distant stars does not bend due to force of gravity from other stars, and can be validated by taking pictures of the stars from an astronomical laboratory located on the surface of the moon or from the International Space Station. We are sure that NASA and ESRO could easily figure out the details of such measurements and experiments.

We want to make some suggestions for the space station program. The space agency should design and deploy an emergency escape vehicle that would be able to safely carry the astronauts from the space station down to the Earth, in case of a real emergency. The vehicle's main payload should be the weight of the astronauts. The vehicle could descend to the Earth with a parachute and an alternate source of power, such as a retardation engine. The engine of the parachute should be powered from the space station to steer toward the Earth's gravity. As the vehicle enters the atmosphere, the parachute should open to provide a safe landing. The vehicle should have a GPS so that the ground station can locate its position. Another compact version of the escape system could be used to provide an escape from a space shuttle when Discovery or Atlantis is on a mission to the space station and an emergency occurs during flight.

We would like to make another suggestion to meet nutritional requirement for habitats of space station and in future mission of moon. Occasionally, on television news we hear that the space station ran out of food supply for the astronauts. The space station should be equipped with an emergency supply of food in the form of capsules. These capsules of food could be stored for several months in the space station. In the next section, we shall look at the design considerations for the development of space vehicles with speeds at fractions of the speed of light and in excess of the speed of light **c**.

6.6 Spacecrafts with Speeds above c

For deep space missions, NASA should redesign probes with considerably higher speeds than the manned space shuttle missions such as Columbia and Discovery. The speed of these probes should be as high as the speed of light. The high speed can be attained by ion fueled rocket engines. The task is to design engines that can provide constant accelerations to space vehicles in deep space. There are many other considerations involved. One of the important issue is the design of probes should include retardation engines to decrease the speed of the probe from the speed of light to the controllable range and will allow lending of the probe at destination satellite or planet safely. Also, the onboard instruments should preserve their functions with tremendous changes in the speeds and temperatures.

One of the greatest challenges in the design of vehicles with speeds at the speed of light, or even at fractions of the speed of light, for example, 10% (**0.1c**), is to exchange information from the vehicle to the Earth in real time. Even if hypothetically one could design spacecrafts that travel at or above the speed of light, the speed of communication of information and signals is limited to the ½ the speed of light **c**. The factor ½ comes from satisfying first Nyquist's criteria for zero inter symbol interference and no aliasing requirements on a communication channel [40]. As you may recall, we had introduced the concepts of the absolute speed of light **c** in Chapter 2. Because the speed of any waveform signal can't exceed the speed of light **c** characterized by a unique frame of reference, if the speed of the vehicle is faster than the speed of light, the signal exchange will lag behind in real time. Therefore, it is imperative that all of the computer programs and steps of the mission should be fully autonomous on spacecrafts travelling at or near the speed of light. They should be perfected to the extent that they meet all kinds of emergencies and contingencies without any intervention/assistance from the control center on the Earth. Another issues is tremendous advances in sensor technology is required. At very high speeds various parameter such as spacing, velocity, acceleration and temperature of obstruction entities on the journey should be measured with very small latency. All the sensors systems developed to perform the measurements should have very low value of time constants to report and react to results of measurements. The response of the control systems, designed to alter the directions and change the speeds of vehicles, should be rapid in order to meet the challenge arising from a variety of situations.

One of the major setbacks for Einstein's effort stems from the fact that his work did not address the effect of gravitation on mass of objects. Specifically, his energy equation does not include potential energy stored in matter due to presence of other objects in proximity. Not only Einstein, but several other physicists such as Richard Feynman had decided to ignore the effect of gravity on mass but favored effect of the speed **v** on mass. Several books have derived expression for total mass of object moving at the speed v. The equation is repeated here from equation (4.2):

$$\mathbf{M_v = M_0/ (1\text{-}v^2/c^2)^{1/2}} \qquad \mathbf{(6.1)}$$

where $\mathbf{M_0}$ is the mass of an object at zero speed or rest mass

The energy equation they have applied to find the expression is the time rate of change of energy equal to the product of force and velocity of the object under consideration:

that is, $\mathbf{dE/dt = F \times v}$

This equation was solved for $\mathbf{M_v}$ using the tricks of calculus. The solution for $\mathbf{M_v}$ comes from the equation:

$$\mathbf{M_v^2c^2 = M_v^2v^2 + M_0^2c^2} \qquad \mathbf{(6.2)}$$

However, the results of these computations are based on Einstein's assumption that the total energy of a body always equals $\mathbf{M_vc^2}$. This assumption is not valid when the body is in the neighborhood of other bodies because the energy equation should incorporate the potential energy changes. Einstein himself admitted that he did not know how to incorporate the effects of gravity on mass. According to his work, as the speed of an object increases, the total energy and total mass of the object increase, too. His theory does not answer the question, "why the potential energy stored in the object should not affect the total mass of the object?" It is true the majority of the time, that the object's travel in space does not have a lot of potential energy until the object arrives near other objects, but most objects of interest are molecules that consist of a combination of many particles, not a single particle, such as the neutrino or electron. Also, eventually the fast particle encounters the gravity effects from other objects because they reach the range of other objects when the effects of potential energy are significant to the computation. From our viewpoint, the mass of all objects in the universe is constant unless the atomic configuration of the elements of the object is modified. As explained several times, the mass is a measure of a quantity of the matter contained in the object. By adding kinetic energy to an object, we are not adding material mass. Therefore, the mass change, when the object attains a non zero value for the speed from rest, is not an accurate fact. Also, it is not possible to modify the speed of the object by supplying kinetic energy without affecting the potential energy of the object. Hence, our postulate, that the mass of the object is independent of the velocity of the object, makes perfect sense.

In the next section, we shall analyze NASA's plans for the decade 2010 to 2020. From the economic and mission stand point, this period will be playing a very important role in space and astronomical research for this century, as we are approaching the most critical era of space research. The success of research and missions in the 2020s should help take giant steps for mankind in astronomy and particle physics sciences.

6.7 NASA on the Right Track: Future Missions

In the previous section, we talked about the feasibility of design of spacecraft that may travel near the speed of light in principle. In this section, we shall review the state of the art and some of suggested designs for spacecrafts in missions of 2020's. NASA has announced that they are going to ground the Discovery and Atlantis Shuttle fleet programs in 2012. From 2012 to 2018, NASA, with the help of private industry, will develop new space vehicles. One suggestion is to design separate vehicles, one for carrying astronauts to their mission destination site and another for carrying cargo payload. We believe that this is a step in right direction. There are many advantages of employing different designs, customized for each purpose. One of the most important advantages is that space vehicles, carrying astronauts only, could be designed to fly considerably faster than the speed of the cargo missions. Thus, travel time for astronauts to mission destinations could be a lot shorter than slow missions with cargo. The efficiency of the implementation can be dramatically increased because the life support system payload requirements during travel is reduced, which may result in a huge cost savings.

Another advantage of the mission, carrying the astronaut's payload only, is that the space vehicles could be designed with increased safety at a low cost, thus mitigating risk. A different design for cargo missions will demand that the industry develop space vehicles that are capable of running self-directed missions with completely autonomous control systems. Further missions that carry cargo payload may not always require the roundtrip fuel and resources because they could be optimized for one-way trips sometimes. The entire mission operation and control may be customized for each purpose with very high efficiency. Further, this will increase research demands in public and private sectors and help stimulate economy by creating millions of new jobs domestically and world-wide. We are hoping that current administration will realize that timely introduction of this book to students and practicing professional will gain a great deal of insight to promote their careers. After looking at the success that NASA has achieved for missions at International Space Stations, we are sure that the NASA is on the right track. Also, grounding the Atlantis and Discovery space fleet will relieve NASA personnel and free them to think and implement future space missions with fresh ideas.

6.8 Summary

In this chapter our focus was to explain that there is no need to invest funds to build particle accelerators and particle colliders. We showed that single particle proton or neutron can be accelerated to the speed of light if occurrence of emission event for radiation energy is prevented. The main reason the proton or neutron fails to attain the speed of light, is that when its speed approaches the speed of light, the probability of a radiation event increases with the probability of collision. At that speed, the particle collides with another particle, emits radiant energy and results in the loss of the kinetic energy of the particle. Further, even if a single particle proton or neutron attain the speed of light does not imply that particle consists of molecules and atoms could attain the speed of light in particle accelerator. Molecules are formed by combination of several atoms. Each atom is comprised of many protons, neutrons and electrons. The likelihood that the electrons of an atom at that speed will be attaining the excited state and emit radiation, is much higher than if the molecules were travelling at lower speeds.

We concluded that it does not serve any further useful purpose to construct even larger super colliders. Therefore, funds spent toward colliders should be available to perform other research projects, such as to develop very high speed rocket engines. In Section 6.2, we showed that units of measures of time, length and mass remain the same for all inertial systems. Therefore, the Lorentz transformations equations are not required to transform mass, length and time measures among different inertial systems. Next, we explained that it is not necessary to apply time dilation to fast moving neutrino particles for life time computations. The longer life time experienced by the particle is a virtual extension of life time as depicted by slower time clock measured in its frame of reference. We had explained in Chapter 4 that time dilation is fictional event. The event is virtual event because the measured time indicated by claimed identical clock in that frame is not identical to clock in stationary frame. In Section 6.4, we stated that a possibility of developing future weapons systems exist that may utilize principles of smashing protons and neutrons into quarks. If a process with chain reaction to smash protons and neutrons into quarks is discovered it would constitute a major breakthrough in weapon technology. In Section 6.5, we touched base with goals of missions of NASA in the next decade. We explained why it is very important for NASA to develop space vehicles that would travel at fractions of the speed of light. Next, we touched base with issues and major concerns for future space expeditions. We suggested devising a safety vehicle and escape procedure from the space station for emergencies. We suggested the creation of a green house and laboratory on the moon for more accurate mapping of the sky and to perform a variety of measurements of celestial objects.

In Section 6.6, we discussed some of the factors that affect the design and production of space vehicles with velocity at or near the speed of light. Next, we stated various design considerations to increase the speed of space vehicles to fractions of the speed of light. We emphasize the development of sensor technology, to create sensors with a very short latency to perform the speed measurements, the time of approach to the obstruction, and the temperature and distance measurements. We also explained that control systems and command systems should be fully autonomous because real time control signal and information transfer is not feasible at the speed of travel. The information propagation delay from the control tower on the Earth to the target systems on the spacecrafts is much higher than required by the sensor systems. Another requirement is all the sensors and measuring instruments on board should sustain their operations during acceleration stage. Therefore, the design of these instruments should be validated and tested on the Earth with very high stress limits and margins.

In the next chapter we shall discuss the concepts of Universal Unified Field Theory (UUFT). For many years scientist were puzzled about integration of force of gravity into standard model. We shall show that it is possible to integrate gravity into standard model. Also, we shall discuss the reasons for existence of force of gravity.

7.

Universal Unified Field Theory

In this chapter, we shall apply the concepts of the **Skylativity®** theory to integrate the effects of the force of gravity in the standard model. As a result, we shall develop the UUFT that consistently makes sense between Newtonian mechanics and quantum mechanics, without the need for the intermediate Einstein theory of relativity. In Section 7.1, we will describe the details of the standard models to explain the fact that it makes complete sense for the current model of the atomic structure, as described by Bohr and others. The weakest force of the Earth's gravity on subatomic particles, protons; neutron and electrons, has a very small effect compared to the strong forces of the electric charge and centripetal force, due to the rotational kinetic energy of the electrons in orbit. The ratio of the force of gravity compared to the forces of the electric charge and magnetic spin moment is as small as $\mathbf{10^{-42}}$.

In Section 7.2, we shall attempt to answer the basic question about why the force of gravity exists among different objects. We will prove that the loss of the mass of the objects, because of the propagation of the gravitational energy from celestial or any object is an incorrect principle. It is not correct to say that the force of gravity propagates by the movement of fictitious particles gravitons. The situation is analogous to the photon particle's description in light waves. Next, we will explain that the orbits of the electron in atoms need not be modeled as a circle or ellipse, as proposed by standard model. It is possible that the electron orbits may be more accurately modeled as complex sinusoid instead of orbits similar to the planets in the solar system. Frequently, the standard model has predicted that the electrons orbit as a cloud and their position is characterized by the colors, that is, a description of a probability that an electron may be found in the position. Coincidently, the color names are also used to distinguish some of the quarks.

In Section 7.3, we will try to answer the question of time travel. Even though time travel is an interesting thought, it is an imaginary fact. We will discover that time travel is not feasible for humans. In the strictest sense, all of the spiritual processes are irreversible and inhibit time travel reality. In Section 7.4, we shall try to explain why the weather and climate forecast frequently fail. We decided to include this in our discussion because it is of interest to many students in the physics community.

7.1 UUFT: Integration of Gravity in the Standard Model

Another significance of the new theory postulated in this effort is it solves the issue of unified field theory. For many years scientist had trouble applying Einstein's theory of relativity to quantum mechanics (wave theory) and Newton's theory of gravity (Classical theory) tying them together with logic and consistency. The novel theory makes perfect sense. We like to call it Universal Unified Field Theory (UUFT). Einstein's theory of relativity did not answer the following question pertain to nuclear model of atoms from Ernest Rutherford. What keeps the negatively charged electrons from falling into a positively charged nucleus by the electro-static force of attraction? The nucleus of atoms consists of protons and neutrons. What prevents exploding of protons in multi-proton atoms on account of the repulsion of positive charges?

The answer to the first question came from Niel Bohr. He figured that electrons were orbiting surrounding the nucleus similar to the planets orbiting around the sun. He stated that instead of the gravitational force in effect in the planetary system, the electric attractive forces between the positively charged nucleus and the negatively charged electrons, would supply (energy) the centripetal force required to preserve the orbit of electrons. One concern was that the revolving electrons will lose energy by radiation and eventually collapse with nucleus. However, Niel's postulate, that the electrons in atoms exist only in any one of a number of stationary states in which no emission of radiation takes place [22].

Any absorption and emission of light or electromagnetic radiation will correspond to a sudden discontinuous transition between two such states. Planck suggested that atoms emit light in definite amounts of energy known as quanta. We like to expand the theory further. The orbiting electrons are balancing the force of attraction between proton and electron, and force of gravity between the nucleus and electron with centripetal force. The centripetal force is due to the kinetic energy of electrons. Thus, the farther the orbit of the electrons, the speed of the electrons is lower than the speed of the electrons in orbit close to the nucleus.

Still, an answer to the second question was not known, that is, why protons do not explode and move away from the nucleus. The answer to this question comes from the invention of quarks as basic building blocks of particles in nucleus namely protons and neutrons. The quark-like particles mesons were first discovered by Japanese scientist Hideki Yukawa in 1949 for which he was awarded a Nobel Prize. Yet the particle was not recognized as a member of family of quarks particles. It was on November 10, 1974, when Burton Richter and Samuel Ting from MIT, verified at SLAC in California that quarks and leptons exists and nucleus of all matters is build from these fundamental particles. Even until the death of Einstein, he was not aware that the core of atoms is made up of quarks. He believed that in the event of nuclear reaction, the energy released by consumed mass was light energy. Somehow mass was translated into light. In reality, radiant energy is released, because of the change in the quantum state of orbiting electrons [22].

The advent of quarks in protons and neutrons explains why protons do not explode. Later it was verified that each proton and neutron consists of three quarks each has a charge of **-1/3 e** or **+2/3 e** where e is a charge on one electron. The force of attraction of positive and negative charges of up and down quarks, between protons and neutrons and neighbor protons balances each other. Soon a standard model for configuration for atoms was developed and widely accepted in physics community [5]. The standard model answered all questions with logic and consistency and satisfied modern scientists completely. It is interesting to note that quark extension to standard model does not explain why masses of bonded nucleons such as protons and neutrons in a helium atom is lower than masses of unbound protons and neutrons. One such instance is the process of the thermonuclear burning of hydrogen into helium occurring interior to the sun. A British astronomer, Robert Atkinson illustrated that this process inside the sun transforms a tiny amount of mass into a large quantity of energy each time the reaction occurs. It was verified that life time of all quarks is very short that they do not exists in nature independently and freely. Therefore, the only way rest mass of matter can be translated into radiation energy is when positrons and electrons are annihilated to produce energy in a thermonuclear fusion process.

Einstein's mass to energy conversion relation estimated total energy released during such a translation. More detailed calculations are required to understand the distribution of radiation energies into energy bands such as infrared (thermal) energy and visible light energy. Obviously, thermal energy determines the temperature of resulting helium and hydrogen isotopes. Typically particles like quarks with larger mass decays or disintegrated into smaller quarks by a process known as annihilation and have relatively short life times. For instance, life time of muon is $\mathbf{2 \times 10^{-6} s}$ and for tau particles is $\mathbf{3 \times 10^{-13} s}$. During the annihilation light waves or radiation energy is released to meet energy conservation.

7.2 Reasons for Gravity

One more question arises to establish the correctness of the unified universal field theory. How to incorporate the effects of the gravity from the huge celestial objects, such as stars, the sun, other planets, and our own planet, Earth, on the electrons, protons and neutrons of the individual atoms of all matter found in the universe? The purpose of the unified field theory is to develop a complete understanding of the interaction of all types of forces involved that govern the state of motion of the most fundamental stable particles, protons, neutrons and electrons. As stated earlier, the stable fundamental particles experiences force of gravitation caused by all objects in the universe, weak and strong forces caused by neighboring atom particles, electrostatic forces due to charge, electromagnetic forces due to magnetic spin momentum vectors and centrifugal force due to orbital motions.

We have a very convincing answer to incorporate the effects of gravity. To simplify computations, consider the instance of forces seen by two orbiting electrons. Those electrons are acted upon by repulsive electro-static force due to negative charge and gravitational attraction. The gravity attractive forces occur due to the mass of two particles. Also, gravity of all other objects including celestial objects acts. The ratio of electrostatic force to gravity between two electrons due to their own mass is $\mathbf{3 \times 10^{42}}$ pg 370 [7]. This implies that the self gravitation force on the electron is very weak compared to the charge force which is an enormously huge number. Further, the gravitational force of attraction between large heavenly bodies and an electron is a very small number. This is true because the gravitation from heavenly object will not be effective in real time, if the objects distance is more than distance corresponds to the distance travelled by light in a second.

From Newton's law of gravitation, the force is:

$$\mathbf{F = (G \times m_e \times m_h) / R^2\ N} \qquad \mathbf{(7.1)}$$

where $\mathbf{m_e}$ is the mass of the electron
$\mathbf{M_h}$ is the mass of a heavenly body
R is the distance between the center of mass of an electron and the heavenly object

For instance, the force from the moon, on an electron residing at the surface of the Earth, will be equal to:

$\mathbf{M_e = 9.1094 \times 10^{-31}\ kg,}$
$\mathbf{M_m = 7.349 \times 10^{22}\ kg,}$

Distance $\mathbf{R = 384400\ km}$
$\mathbf{G = 6.6726 \times 10^{-11}\ m^3\ kg^{-1} s^{-2}}$

Substituting

$$\mathbf{F_{me}} = \mathbf{(6.6726 \times 10^{-11} \times 9.1094 \times 10^{-31} \times 7.349 \times 10^{22}) \div (384400 \times 10^3)^2}$$
$$= \mathbf{446.697 \times 10^{-20} \div 14.776 \times \mathit{10^{16}}}$$
$$= \mathbf{30.23 \times 10^{-36}\ N}$$

The magnitude of electrostatic force of attraction or repulsion between two charges is:

$$\mathbf{F_q = k_e \times q_1 \times q_2 / r^2} \qquad \mathbf{(7.2)}$$

where **q** is the charge on the electron
r is the spacing between the electrons, or a proton and an electron
$\mathbf{k_e}$ is Coloumb's constant

For instance, the force of attraction between an electron and a proton neighbor is:

$\mathbf{k_e = 8.99 \times 10^9\ N\text{-}m^{2/}C^2}$, $\mathbf{q = 1.609 \times 10^{-19}}$ Coulombs and spacing $\mathbf{r = 5.3 \times 10^{-11}\ m}$

Substituting numbers into the equation (7.2) gives:

$$\mathbf{F_q = 8.2 \times 10^{-8}\ N}$$

The ratio of the force of gravity from moon to the force of electrostatic attraction between two electrons

$$F_q/F_{me} = (8.2 \times 10^{-8})/(30.23 \times 10^{-36}) = 0.271 \times 10^{28}$$ which is very large number.

Also, the upper limit on the difference of the gravity force between two electrons and a heavenly object is proportional to the distance between two electrons:

$$\mathbf{F_{e1}} = (\mathbf{G} \times \mathbf{m_e} \times \mathbf{m_h}) / \mathbf{R^2}\ \mathbf{N} = (\mathbf{G} \times \mathbf{m_e} \times \mathbf{R} \times \mathbf{m_h}) / \mathbf{R^3}$$
$$= (3/4)\ \mathbf{G} \times \mathbf{m_e} \times \mathbf{R} \times \rho/\pi$$

$$\mathbf{F_{e2}} = (\mathbf{G} \times \mathbf{m_e} \times \mathbf{m_h}) / \mathbf{R^2}\ \mathbf{N} = (\mathbf{G} \times \mathbf{m_e} \times (\mathbf{R+r}) \times \mathbf{m_h}) / \mathbf{R^3}$$
$$= (3/4)\ \mathbf{G} \times \mathbf{m_e} \times (\mathbf{R+r}) \times \rho/\pi$$

the difference is $\Delta \mathbf{F} = \mathbf{F_{e2}} - \mathbf{F_{e1}}$

$$\Delta \mathbf{F} = (3/4)\ \mathbf{G} \times \mathbf{m_e} \times \mathbf{r} \times \rho/\pi \qquad (7.3)$$

where **r** is the spacing between electrons
and ρ is the density of the heavenly object

For $\mathbf{m_e} = 9.1094 \times 10^{-31}$ **kg**, $\mathbf{r} = 5.3 \times 10^{-11}$ **m** and $\mathbf{G} = 6.6726 \times 10^{-11}\ \mathbf{m^3\,kg^{-1}s^{-2}}$

$$\Delta \mathbf{F} = (3/4) \times 6.6726 \times 10^{-11} \times 9.1094 \times 10^{-31} \times 5.3 \times 10^{-11} \times \rho/\pi = 83.569\rho \times 10^{-53}$$

Again the ratio of electrostatic force and this gravitational component is greater than $\mathbf{3 \times 10^{42}}$. Therefore, the gravitation is much too small to have any effect on the cohesion of the electrons. Also, according to the quantum theory, the position of an electron can only be specified as a probability density function, a cloud. At this high speed near the speed of light the effect of gravitation on nucleus particles due to heavenly objects cannot be precisely determined due to the Heisenberg uncertainty principle. The high speed of the electrons predicted by the standard model is consistent with the current conduction rate in MOS switches for the circuits fabricated on a semiconductor wafer. The current flow is proportional to the speed of minority carriers in semiconductor material. This proves that gravitational force from large heavenly bodies is of no consequence inside lattice structure of atoms.

Yet another dilemma faced by modern scientists is they have failed to answer two more questions. Why there is a gravitation force and interaction exists between any two bodies in our universe? The second question is how the effect of gravity force field propagates in empty space without any medium and at what the speed? The first question may be answered as follows. Some properties of particles found in nature are fundamental properties to the particle. For instance, charged particles (both positive and negative charges) create an electro-static field in the vicinity. Particles with magnetic dipole alignment develop a magnetic field in the proximity of a pair of particles. Particle vibrating at infrared frequencies radiates thermal energy. Particle releasing vibration energy with frequency corresponds to visible light develops a luminous field. The field illuminates the space when light wave propagates through the space. Similarly, force of gravity exists between two particles by the fundamental property.

The fundamental property that governs the gravitational force arises from fact that every object possesses moment of rotational inertia. Because of the dynamics of the object (the movement of molecules, atoms, and fundamental particles), the object possesses rotational inertia. This rotational inertia provides for gravitational energy and force from the object on other objects. The fundamental property is developed by the fact that principle of conservation of energy holds well for any and every frame of reference. From Bohr's atomic model it is obvious that every particle found in nature possesses kinetic energy. To balance kinetic energy for a neutral mass there has to be a potential energy associated with the mass and another body. To attain a stable position for first mass in relation to second mass, the kinetic energy should be balanced by the potential energy. The potential energy is supplied by force of gravity exerted by the second body on the first body. Therefore, force of gravity comes into play between two bodies or particles found in nature. It is well known that light waves does not require medium of propagation. Similarly, the force of gravity between two bodies does not require a medium to propagate. As regards to the speed of propagation of force of gravity, we believe that it is variable and depends on frame of reference. The maximum value for the speed at which force of gravity propagates is the same as the true speed of light in the frame of reference as described in Chapter 2.

To explain the wide spectrum of radiant form of energy from radio-active elements Dr. Greene [14] developed concept of Strings. He described that fundamental particles of neutron and protons; quarks consist of loops of strings. One of the primary limitation of his concept is validation of these strings is unrealistically difficult. The life time of quarks is very short (few tens of micro-seconds). Therefore, existence and verification of presence of strings in quarks will involve development of advances in measurement techniques which are too far away in terms of time-frame. Instead, we believe that the event of the different radiation spectra from the atoms of the matter in an active state is best described by the motion of the orbiting electrons. The orbit of the electrons surrounding the core may not be an exact circle or ellipse like the orbits of the solar planets. The orbit of the electrons may be modeled as a complex sinusoid with period $\mathbf{2 \times \pi \times f}$ and the average position, a circle on the surface or boundary of the sphere of the energy shell for the specified electron state. Figure 7.1 illustrates the electron orbits inside an atom surrounds the core of the nucleus at microscopic scale. The trajectories of the electron orbits may actually be sinusoidal with the center line of an ellipse, instead of an elliptical or circular orbit shell. Also, the orbits of the electrons in the outer shells are not in the same plane as the plane of the electrons in the inner shells. This may be one of the reasons that the electron orbits are found and characterized by different colors in the standard model for atoms. Here, the color means the probability of finding the orbiting electron in a specific position.

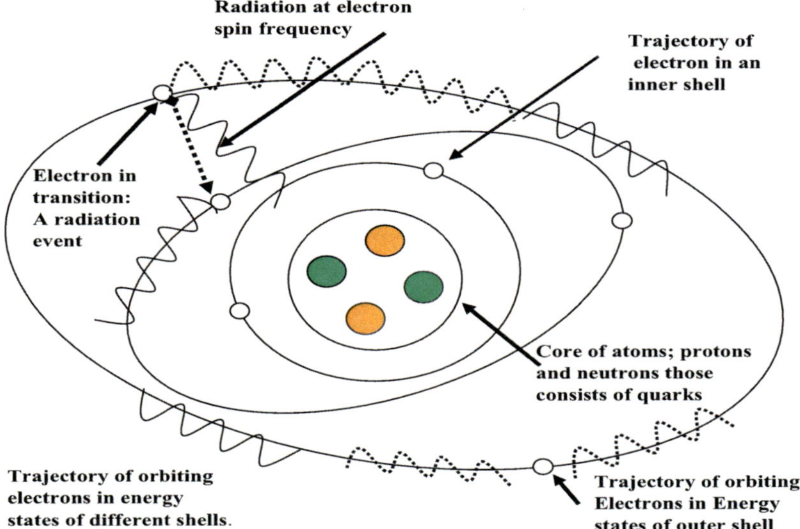

Figure 7.1 Structure of atoms: Strings beyond quarks. Since electrons are always in motion inside atoms and they change orbits according to probability of changes in quantum state, it is possible that the electron orbits may resemble a string like configuration. We predict an electron's instantaneous positions may be described by the infinite Fourier transform series.

Next, we will see the reasoning that the electron orbits may not be circular, makes perfect sense. The orbiting electrons are under the influence of several force vectors, such as the electro-static due to charge, the electromagnetic caused by the spin momentum, the force of gravity with the core of the atom, the neighboring atoms and electrons, and the force effects of the radiation fields, such as light or infrared, when the electron is in excited state. The vector summation of the momentum caused by these forces determines the instantaneous position of the electron in orbit. The precise determination for the trajectory of non-circular orbits for the electron will allow an explanation of the relationships between the frequency and the radiation energy with very high accuracy when the energy is released from an electron in an excited radiant state. According to Louis de Broglie's hypothesis, the wave-particle duality is a general property of both radiation and matter. Many experiments were performed to verify the fact.

According to our view, the wave behavior of the particle electron is a theoretical model. Likewise, the particle photon model for light is a theoretical concept and a tool to facilitate energy, time and other computations. In practice, the electrons are particles and the radiation energies are waves. We believe that the trajectory of the electron orbit is a wave and not just a circle on a sphere or a shell model, as described by Niel Bohr and de Broglie. The wave model for the orbits of the electron is more closely related to the string theory concepts introduced by Dr. Brian Green [14]. In his national bestselling book, the Elegant Universe, which had won a major NOVA special on PBS, he described that the subatomic particles electrons and quarks are fabricated by a tiny loop of vibrating strings. Since it is not possible to analyze the structure of quarks beyond the description of quarks as charge particles with charge **1/3q** units, it is less probable that quarks may consist of loops of tiny strings. Further, the life time of quarks has been found to be too short to perform any measurements. Therefore, it may suffice that the string type of motion of the electrons in orbit, approximates the concepts of the string theory developed by Dr. Greene. One should keep in mind that the electron orbits are complex sinusoids; the specification for the light energy released and modeled by Planck as a product of $\mathbf{h} \times \mathbf{v}$, needs to be characterized with high accuracy.

As explained earlier, the light energy specification and measurement should be related to techniques used to characterize energy of electromagnetic waves. Light is also a form of wave energy. There should not be a significant difference in terms of measurements and specification for both of the energy forms. Our explanation is that the orbit of electrons is a complex sinusoid is consistent because it is well-known that the high temperature of atoms is caused and felt as heat when the electron vibration amplitude is increased. The tasks to verify that the electron orbit is not circular or elliptical, instead is a sinusoid, is left to physics students as an exercise. In the next section, we shall discuss a topic that is of interest to everyone who grows to understand the concepts of time and time travel. From the time that stories about time travel were written by Jules Verne, several movies were produced to describe strange circumstances that arise from time travel using time machines. We shall see that the time travel concept exists in fiction movies only.

7.3 Time Travel

Many scientists for several years have been intrigued about time travel. We are able to travel back and forth in space in all three dimensions **X**, **Y** and **Z**. Meaning, we can move left or right in **X** dimension, or X-axis, coordinate values positive or negative with respect to a center. We can move forward or backward in **Y** dimension, or Y-axis. Similarly one can freely move upward or downward from a center origin **Z** dimension, or Z-axis. A good question is can we move in fourth dimension T time axis. Answer is no. No matter what we do or make change to state of motion of objects. We can never change the rate at which value of time that is measured. Absolute measurement of time is not modifiable quantity. Time is invariant with respect to space, time or any frame of reference. One can neither slow down time nor make it tick faster with any degree of knowledge and or certainty. Let us look at few ways in which standard time clocks are designed. Also, we shall look at the meaning of standard time.

Historically, people believed that time, as defined by the orbital period of the Earth and the other planets surrounding the sun, should be constant. It turns out that indeed the rate is fairly constant because mass of the sun is constant in-spite of radiation energy loss from the sun. The main reason is that number of baryons (nucleons, sum of protons and neutrons) on the sun is constant as per the baryon number rule. Even the total number of electrons on the sun is not alterable. According to the postulates of the theory of **Skylativity®**, light and radiation energy is emitted from the sun as a result of the changes in the potential and kinetic energy of electrons. No mass is converted into light waves or consumed. Therefore, the size of the solar year has not changed by even the most miniscule fractions of a second over the past several hundreds of millions of years. Therefore, clocks designed on the basis of the Earth's periodic motion surrounding the sun should provide a good and highly accurate measurement of time.

This is the greatest benefit from nature to mankind, that the time rate is preserved. Even if the core temperature and the inner core of the sun, is cooling off and slowly transformed into liquid and solid, the mass of the sun is not changing. According to Dr. Roger Freedman and Dr. Walter Kaufmann III **0.7%** of mass is converted into radiation energy from every four atoms on the sun [17]. This is a very high rate of loss of mass. Therefore, we believe that the energy is not radiated because of the conversion of two electrons into radiation energy which is released during the thermonuclear burning of hydrogen into helium. We believe that the thermonuclear burning of hydrogen atoms results in the stable helium gas atoms. The fact that the orbital period of the Earth's rotation surrounding the sun is constant and verified by other standard of time such as atomic clock proves that the mass of the sun has not changed. This fact substantiates our postulates that the source of light and the thermal energy during nuclear or any radiation energy, is by changes in the potential energy of electrons in the orbits of atoms involved in the reaction and not the actual burning of nucleons into energy waves.

It is predicted by Chandrasekhar [26], that after 4.5B years the sun will be cooled down and solidified into a white dwarf star. Also, it is predicted that the sun was formed 4.5B years ago. The transformation of $\mathbf{H_2}$ into **He** gas should result in a cooler core of the sun than predicted. The temperature profile of the sun's core, should be a complete bell shape, different from what we had discussed in Section 2.7, and should not look like a monotonically increase to value **15 M°K** at the center of the sun [17].

Next, we will look at **Cs** atomic clock time standard. The National Institute for Standards of Technology maintains a standard of time source as an atomic clock. The atomic clock provides a very accurate time measure. The frequency of the spectral line emitted from the **Cs** compound is considered highly stable. The stability of the clock is 1 part in $1B^{th}$ of a second. The amazing fact is that the period of rotation of the Earth does not change by even a fraction of a second. If the sun was losing mass, the period should change significantly and should be noticeable. To complete our discussion on time travel, let us stipulate the reasons why time travel is not realistic.

As explained, it is not possible to change the rate at which time elapses. Secondly we know that spiritual processes are irreversible. The issue arises from the fact that when a person or human or any animal dies there is no means were the person or life can be brought back alive in the physical body. Time travel demands that spiritual processes should be reversible to replicate events of past tense or to create future events at a space coordinate. These conflicts cannot be overcome. Therefore, time travel is impossible and it can only happen in fiction. In the section 7.4, we shall address a slightly different topic, the accurate prediction of the weather because it may affect the life style of human civilization to a much larger extent than most of us realize. In particular, with the advances made in technology, significant progress has been achieved in forecasting the weather with tremendous accuracy, yet the challenge is not complete.

7.4 Weather Forecasts

One of the most difficult problems in everyday life that concerns all of us is accurate prediction the weather. With advances in technologies, computer simulation and models based on real-time data available meteorologist does very good job of prediction. However, sometimes their prediction fails. We like to explain why it is not possible to accurately predict the weather at all geographic locations with 100% accuracy. The weather changes occur because of combination of many effects. The majority of these effects are predictable very systematically. The systematic effects are the movement of the Earth surrounding its axis, the rotation of the Earth in its orbit surrounding the sun, the movement of the water front in the ocean tides, and the varying degree of vapor pressure, due to differences in the evaporation rate, altitude, etc. The main cause of weather change is that they are not predictable all of the time because the motion of the gas molecules of the atmosphere is random. The motion of gas molecules is caused by temperature which is affected by enthalpy. From statistics it is known that outcome of random event can be predicted with 100% accuracy only on average basis and not instantaneously. Most of the times effect of random motion of molecules on overall weather condition may not be significantly high. When weather condition is disturbed by dominant effect of one or more systematic cause create a situation thereby the random motion of molecules effects will be more pronounced in local regions. The disturbance is magnified to the extent that the predicted weather may be wrong in those situations.

7.5 Summary

In Section 7.1, we showed that the effect of the force of gravity on subatomic particles can be successfully integrated into the standard model. It turns out that for the size of subatomic particles, the force of gravity has a very weak effect compared to the strong effect of forces due to electric charge and the forces due to magnetic spin momentum. In Section 7.2, we explained the reasons why the force of gravity exists in nature and is portrayed by every object. Several modern physicists, such as John Hutchinson and others have developed techniques that demonstrate the possibility of objects that possess the property of anti-gravity for a controlled duration of time with limited success. This is still research in progress.

In Section 7.3, we talked about the issue of time travel. We concluded that time travel is not possible. We substantiated our argument by stating that spiritual processes are irreversible and one cannot alter the rate at which time elapses. We proved that the mass of the sun remains constant in-spite of loss of radiation energy. Our conclusion is based on the fact that the number of seconds in the period of the Earth's orbit surrounding the sun has not changed to any measureable extent for the past millions of years. Finally, we explained why it is not feasible to forecast the weather with 100% accuracy at all the times and at all locations.

The next chapter is about black-holes, supergravity, the size of our universe, the classification of the universe, the origin of the universe, the fate of the solar system and our civilization, and the retrograde orientation of rotation for some of the planets. Also, we shall touch on the concepts of modern physics theories, string theory, and new dimensions.

8.

Black Holes and Infinite Universe

In this chapter, our focus will be to explain black holes. In Section 8.1, we shall explore why light and other radiation energy waves seemingly do not return when directed toward black-holes. We shall show that the claims of Steven Hawking and others, about supergravity possessed by black holes that traps the light waves, do not make sense. In Section 8.2, we shall discuss the theories behind the creation of our universe. We shall show that it is more probable that the universe was created as a result of more than one big-bang event as opposed to the current theory of a single big-bang explosion. We shall prove that the universe is neither contracting nor expanding, as claimed by Hubble and others. In Section 8.3, we shall demonstrate that it is not sufficient to describe the Hubble Constant in one dimension. Also, we will describe issue of stability in universe. To explain stability, we will state and prove a theorem that is obeyed by the motion of all celestial objects. In Section 8.4, we shall examine the limits of observation within the universe. It turns out that only a very small section of the universe in our proximity can be observed in real-time. The rest of the universe is observable in the past time. Therefore, we suggest the maps of the universe should be partitioned in three segments: past, present, and future. We shall discuss the various observable partitions, based on information we receive from those parts of the universes. In Section 8.5, we will discuss the ultimate fate of planets in solar system and the star Sun. In the section, we will emphasize a fact large fraction of mass active stars such as sun is conserved during the course of its main cycle life span. Also, we will state and prove mass conservation in Universe theorem. In the same section, we will develop technique that will allow us to determine life span of our civilization and the sun with much higher accuracy than before.

In section 8.6 we will explain the cause for east to the west axial rotation of some members of planets in our solar system the Venus, the Uranus, and the Pluto. In those planets the sun rises in the west and sets in the east. In section 8.7, we will describe long elongated orbit motion of comets. We explain the reasons for such oblong orbits of various comets including Halley's Comet. In section 8.8, we will discuss string theory created by Dr. Brian Green and show that the ideas developed by him may lead to creation of theory of everything.

8.1 Black Holes and Fictional Singularity

In the early 1970s, the existence of black holes were first verified when scientists received an X-ray burst from the Cygnus-1 binary system of stars. The burst of X-rays were flickering at time scales of **1/100th** of a second [17]. Since then, a lot of research has been done to detect the variety of black holes found in the universe. Black holes are classified into three major groups: super-massive with masses in excess of $\mathbf{10^6}$ to $\mathbf{10^9}$ solar masses found at the center of galaxies, mid-massive size of **500** solar mass such as M82, and the primordial black holes the size of the Earth to as small as **5 × 10^{-8} kg**, as proposed by Steven Hawking from Cambridge University. Of these three types, the existence of primordial black holes is not verified in nature. For large size black holes, three numbers completely describe the structure of a black hole: its mass, the total electric charge, and its angular momentum. Scientists claim that at the time of the formation of black holes, the electric charge on the black holes would disappear by process of neutralization. They also claim that black holes are free of a magnetic field. These assumptions are not correct. In the following discussion, we shall attempt to answer the following questions as they relate to black holes with massive gravitational forces that even trap light waves. Though current theory, which claims that black holes have a supergravity field, explains the accretion disk formed surrounding the black holes as matter from neighboring stars is pulled, it does not explain the following mysterious questions:

a. Why the trajectory of light waves just outside the event horizon is not circular?
b. How the matter inside the black holes has density higher than the densest mineral found on the Earth?
c. How could the self gravity force of black holes overcome strong forces, such as residual and electric charges, to compress neutrons and protons beyond current levels of density? We shall see that black holes should meet this requirement to support current theory.
d. Why black holes are charge neutral?
e. How the black holes will evaporate if light from its surface cannot escape?

Now we shall shift our attention to the main subject of our discussion, that massive black holes do not have super-massive gravity. Many scientists claim that light rays incident on black holes are not reflected because they are trapped by massive gravitational pull. We do not think that it is the reason for no reflection of light. We believe that light is not reflected from black holes because it is absorbed; similar to light inside a coal mine will be absorbed. It does not make sense that the density of massive black holes is so high that even light cannot escape. We state this because the upper limit on density in our universe is driven by the density of the fundamental particles protons and neutrons. No object in the universe has higher density than the density of a proton or a neutron. Suppose we form an object that is hypothetically comprised of closely packed neutrons. Now, if we target a beam of light on the object, does the object trap the light and not allow it to escape? The answer to this question is that light shall be reflected and not trapped. This means that black holes do not have super density and the gravitational pull to trap light waves. This argument comes from the fact that neutrons and protons cannot be further compressed because they are fundamental particles, and they cannot be affected by a weak force such as gravity. To avoid vigorous computations, we shall provide an indirect proof of our proposed explanation that there is no reflection of light from black holes. Here we will take one step further and provide a formal proof for the questions raised by the Magueijo, about black holes. What if black holes aren't really holes after all? [16].

Though no one has discovered the internal composition of black holes, it is highly likely that they should be comprised of some kind of gas or solid matter. To support the present claim that black holes are massive and possess supergravity implies that they are made up of solid matter. All matter found in nature so far, consists of a combination of the fundamental particles, protons, neutrons and electrons. It is safe to assume that the matter on black holes is also composed of molecules and atoms that are in turn comprised of fundamental particles. Since fundamental particles, neutrons and protons are comprised of the sub nucleus particles, up and down quarks, they are held together by the force of the charge from the quarks. Our premise is that the fundamental particles, protons and neutrons, cannot be compressed further by the gravitational forces which are proven to be considerably weaker than the charge forces for sub nucleonic particle sizes. Inside the black hole, the fundamental particles of matter are likely to be affected by three types of forces. Namely, the force of the gravitation of the massive black hole, the force because of the charges on quarks, and the compression or expansion force caused by thermal effects. If one analyzes the relative strength of the forces of interaction (refer to Table 2.5), they can conclude that the neutrons and protons inside the atoms of constituent elements of matter on the black holes, cannot be compressed further. Therefore, the density of matter inside the black holes cannot exceed the density of the fundamental particles, neutrons and protons. Let us construct a hypothetical object which contains tightly packed neutrons only. We will call this object a neutron star. We will study the escape velocity of the point particle from the surface of this super dense star. If we prove that the density of this neutron star is such that it will prevent the escape of light waves from the event horizon distance then, we will call it a black hole. The critical radius, at which the escape the speed is **c**, is called the Schwarzschild radius. To facilitate our computations, we will determine the largest size of the densest neutron star from which light waves cannot escape.

First, we compute the density of the hypothetical neutron star as follows:

$$\begin{aligned} \text{Density of a neutron star } \rho_n &= \mathbf{M_n/V} \\ &= \mathbf{(M_n/R^3_n) \times (3/4\pi)} \\ &= \mathbf{(1.67 \times 10^{-27}/(1.2 \times 10^{-15})^3) \times (3/4\pi)} \\ &= \mathbf{2.3 \times 10^{17}\ kg/m^3} \end{aligned} \qquad \mathbf{(8.1)}$$

where $\mathbf{M_n}$ is the mass of the neutron and $\mathbf{V}$ is the volume of the neutron

Next, we shall use the formulae for the escape velocity, from text book by Serway and Jewett [22], to compute the largest radius of the neutron star from whose surface, light cannot escape:

$$\mathbf{V_{esc} = ((2G \times M_n)/R_n)^{1/2}} \qquad \mathbf{(8.2)}$$

We shall solve for $\mathbf{R_n}$ after expressing $\mathbf{M_n}$ into density and using **c** the speed of light for $\mathbf{V_{esc}}$. In that case the previous equation transforms to:

$$\mathbf{R^2_n = [\ (3/8\pi) \times (c^2/(\rho_n \times G))]} \qquad \mathbf{(8.3)}$$

Substituting values $\mathbf{c = 2.99792458 \times 10^8\ ms^{-1}}$, $\mathbf{\rho_n = 2.3 \times 10^{17}\ kgm^{-3}}$ and $\mathbf{G = 6.6726 \times 10^{-11}\ m^3kg^{-1}s^{-2}}$, to solve for $\mathbf{R_n}$ we get:

$$\mathbf{R_n == [\ (3/(8 \times 3.142)) \times (2.998 \times 10^8)^{2/}(2.3 \times 10^{17} \times 6.673 \times 10^{-11})]^{1/2}}$$
$$\mathbf{== 2.644 \times 10^4 \text{ m or } 26.44 \text{ km}}$$

which translates to 16.42 miles, an astonishing fact!

From the previous computations, it is evident that light will escape from the surface of a black hole constructed from a neutron star whose density is the highest possible density if its radius is greater than 16.42 miles.

In reality, it is believed that the size of the black holes are much larger than 2-3M solar masses and it is not hard to predict that the density of matter inside the black hole should be lower than the upper limit density of the neutrons. Thus, we have proven that light can escape from the majority of the black holes. The only way light is not reflected from the surface of the black holes is by the complete absorption of light.

These black holes are massive yet charge less, this fact is quite conflicting. In order for these black holes to be charge free, it is not clear from where the electrons in the universe will be dragged toward the black holes to neutralize the positive charge on the black holes. Further, the electrons cannot be dragged outside the event horizon for the black holes. Therefore, scientists predict that a singularity that exists at the center of black holes is not accurate. We believe the singularity at the center of the black holes is a fictional concept. Further, scientists also believe that the black holes evaporate. If the black holes possess super gravity that will not allow escaping light waves, how it possible is that it will lose its mass by the conversion of gravitational energy into the particles that quantum mechanically leak out. In order for black holes to lose mass at any rate, an enormously large number of particles should be leaking out from the event horizon. There is no such evidence that the particle escape rate is detected from any of the massive black holes. Singularity at a center of the massive black holes is fictional and it does not make sense to speculate the nature of singularity when the distances of these black holes are unreachable with the current means of travel. The mass of black holes cannot decrease because of the energy loss through the radiation of gravity waves. The gravity waves do not carry any mass similar to light waves. We demonstrated that light waves are created as a result of the change in the quantum energy state of the orbiting electrons in the atoms of the source of light object. The force of gravity comes into play between any two objects when they are in proximity. The gravitational force only exists if two objects are involved. Therefore, no energy or mass is transferred or lost from any object when the force of gravity is felt.

In Section 8.2, we shall discuss the events that would have lead to the creation of our universe. Specifically, we will find out that our universe is the way it is now as a result of several events similar to an event currently proposed by the big-bang theory.

8.2 Infinite Universe with Multiple Big-Bang Events

The science that governs the creation and fate of the universe is known as cosmology. As per the present state of art, it is believed that the universe was formed as a result of a single big-bang event. Also, it is believed that since the time our universe was created, it has been going through an expansion phase. This means our universe is continuously expanding. This fact was discovered by Hubble the first time. He noticed that the red shift of light arriving from distant galaxies indicates that they are moving away**. The rate of recession or z-shift is proportional to the velocity of recession.** He summarized his findings into his law and expressed it in a relationship:

$$\mathbf{v} = \mathbf{H_0} \times \mathbf{d} \quad \mathbf{(8.4)}$$

where **v** is the recessional speed of the galaxy, **d** is distance
H_0 is Hubble's constant

We firmly believe that the fabric of space itself cannot contract or expand, nor can it move. Only the matter or object within space can change their volume or position. Therefore, Hubble's law requires modification. He should state that the galaxies are moving away from each other and away from our solar system. Hubble explained the expansion of the universe by comparing an analogy of expanding a balloon. He said, the size of the coins on a balloon do not expand. Similarly, objects within the galaxies do not expand. This adds error to his proposed law. Actually, the coins will be expanding, too, if one blows the balloon and examines the expansion of the surface of the balloon closely. We find that the fabric of the space within our galaxy does not seem to expand to any appreciable extent. A side effect of Hubble's law is that the cosmologist believes that the universe was formulated from a point and it expanded into multiple celestial objects by a single big-bang explosion event.

We have a different school of thought, when it comes to the origin and creation of the universe. We believe that the space in the universe always exists because the expansion of the space of the universe assumes that there was space into which the space of the universe is expanding. At the beginning, the universe was unbound, untimed and infinite. Meaning, the universe has no boundary and there is no known time when the universe came into existence. As time elapsed, the various celestial objects in the universe were formed depending on the suitable prevailing circumstances. Further, we believe that the distant galaxies, black holes, nebulae, quasars, and supernova were born because of the multiple explosions and not one big-bang event. As a result of one of the big-bang events, our solar system formed approximately 4.5B years ago. Because of the uncertain origin and creation of the universe, does it make sense to ask the question about when the universe will end? No, because the sun will die much sooner than the end of the universe. It is estimated that our solar system will collapse when the sun cools off and becomes a cold white dwarf star. This will happen after approximately 4.5B years. Our belief is that even after the death of our sun and the solar system, the universe will continue to exist. Therefore, it may be imperative to invent systems that will allow prolonging and sustaining life on Earth as it exists today and in the distant future.

We believe that the big–bang theory for the creation of the universe raises many controversial questions that contradict themselves. It suffers from an unanswered paradox of what we describe as thermal inertia discrepancy. For instance, the time lines, temperature, and energy of a particle, after the creation of the early universe in Planck time, predict the unrealistic cool down profile for the universe. In Table 8.1, we have described time lines, temperature, and particle energies, as projected by the grand unified theory (GUT) after Planck time, in the early universe after the big-bang event [17]. The GUT of forces is also known as the theory of everything as it claims to unite the effects of strong, electromagnetic, weak, and gravitational forces. From the table data in the temperature column, it is evident that it would require almost zero time constant and zero thermal inertia for the universe in order for the temperature to drop at those values. It is not obvious that at those particle energies how the heat among the particles will be exchanged to realize the drop in temperature. Further, the energy from the universe was depleted or dissipated into the environment as a result of the loss of temperature. There is no account or answer to the energy loss from the early universe in the GUT. As we will see later, it takes almost 4000 years for the light wave energy (photon described by Einstein's theory of relativity), developed as a result of the thermonuclear fusion at the sun's interior, to escape from the sun. The drop in temperature in the universe, predicted by the GUT in the case of the big-bang, does not make sense in that timeframe, if the propagation of light and energy from the sun is occurring at such a slow rate.

Table 8.1 Energy of particles, the early universe as predicted by GUT.

Time after Big-Bang	Temperature of Universe	Energy of Particle
10^{-43} S	10^{32}K	>10^{4} GeV
10^{-35} S	10^{27}K	10^{4} GeV
10^{-12} S	10^{15}K	10^{2} GeV
10^{-6} S	10^{13}K	1 GeV
2 S		Universe transparent to neutrinos
3 Minute		Helium produced in the Universe
3×10^{5} Years		Universe transparent to light
5×10^{17} S	3 K	At present

In order to determine the age of the universe, scientists measure the change in the frequency of a photon popularly known as red shift **z**. Also, it is determined that the red shift increases as the universe expands and decreases if it undergoes contraction [17]. We agree that the frequency of light would change if the size of universe would change. We disagree that the space within our universe expands or contracts. Only the matter contained in the space will expand or contract, depending on enthalpy (heat content) and entropy (degree of measure of orderliness) changes in the system. This school of thought is consistent with the belief that when any system or components of the system move with a certain speed, it does not imply that the space occupied by the system moves. It does not make sense to say that the space has moved. We would like to describe the effect of the speed of light changes on the distance measurements. Scientists measure the time of flight (round trip) taken by a light wave, which arrived from a distant celestial object in the galaxy, to calculate the distance. If the distant star is receding or approaching at the speed $\mathbf{V_s}$, the relative speed of light, $\mathbf{c+V_s}$ (approaching object) and $\mathbf{c-V_s}$ (receding object), should be used for computations respectively by applying the Galileo transformation instead of using the fixed value **c**. When the distance between the source of light and a moving target is changing, a virtual shift in the frequency of the reflected light signal from the moving target is observed. The apparent shift was first invented by Doppler and corresponds to the speed of the target that is different from a non-moving target. Further, the relative speed of light to the moving target is modified by the same amount.

Another question that often arises, is there the possibility of human life and civilization in existence outside our solar system? We believe that there are approximately **100B** stars in our galaxy, the Milky Way, and it is estimated that there are literally thousands of galaxies in the present day observable universe. We predict that the process of formation and birth of the stars and galaxies in other regions of the universe may be very similar to the formation of our galaxy, the Milky Way, and the solar system. The odds are very high that life would have evolved within the planets, similar to Earth, on other stars, too. Therefore, it is highly likely that the human race may be in existence on another star-planet system. Also, recent advances from the human race in science and technology, increases the odds of the transportation of life to other planets through the contamination processes. Life and living beings flourished on Earth as a result of the natural progression of events in several stages.

Figure 8.1 illustrates the stages in the process cycle that resulted in the successful evolution of life on Earth. As indicated in Figure 8.1a, in the first stage, a small star arrived in proximity of a large star greater than the sun. It is normal that such an event may occur because the universe has estimated over **100B** stars. The probability of this event is somewhat low because the stars are scattered and separated by vast distances. However, as the small star approaches the large star, its speed may increase because of the enormous gravitational pull. In one such incident, the small star collided with a large star and caused a plethora of smashing events and our solar system was created. It is highly probable that this collision event may have elevated the temperature inside the sun's core because of the increased compressive stress. Next, we shall describe how the life force on Earth was originated and eventually populated the Earth with humans and the animal kingdom.

As per the description, in the first stage, the planetary system was born. Figure 8.1b illustrates the sizes of all the planets in the solar system and their average orbit distances from the sun. In the beginning, all of the planets were in gaseous form similar to their parent star, now known as the sun. Since the planets were of a much smaller size than the sun, there was no thermonuclear burning of hydrogen into helium process on the planets that would release energy and maintain a high temperature at their center. Also, planets lost their glow and luminosity after some time, as they started cooling down. More details on the thermonuclear burning, also known as the thermonuclear fusion of hydrogen into helium on the main sequence star, such as the sun, will be provided in Section 8.5. This fusion is an important process which has a very high influence in the determination of the life span of the sun and our solar system. Now, let us focus on the stages of events that lead to the creation of life on Earth.

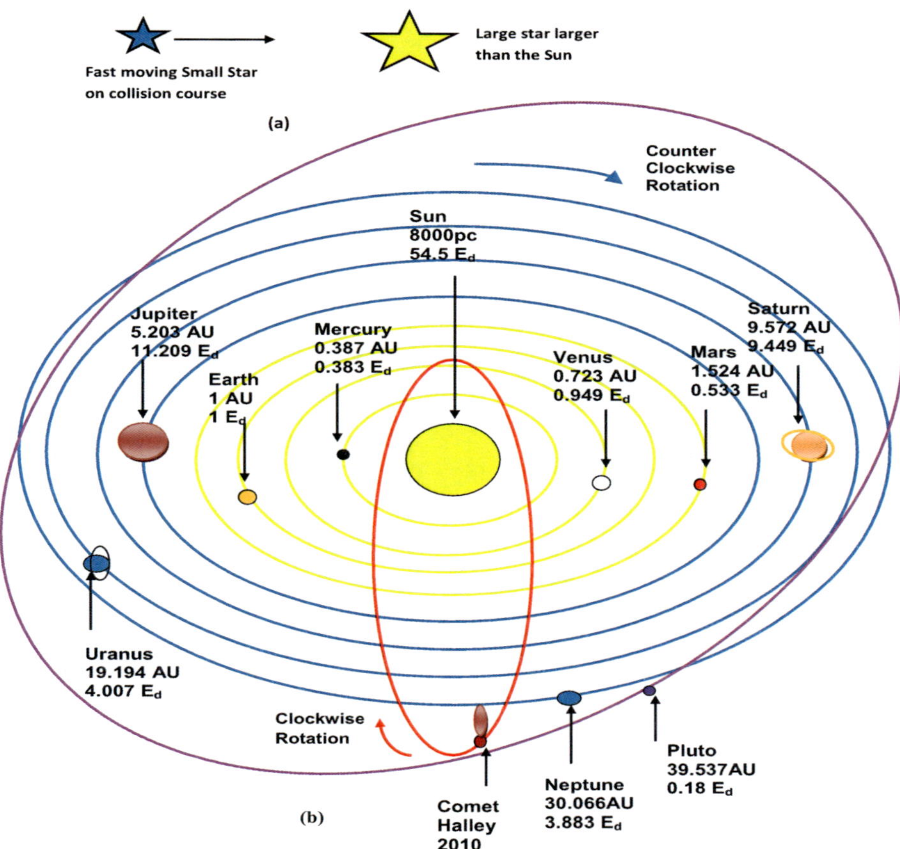

Figure 8.1 Solar Systems. The size of the planets in the Earth's diameter (E_d) and the average orbital distances in the average distance from the Earth to the sun = 1AU (a) The event causing the birth of the solar system (b) The solar system, the distance of the planets from the sun, and the size of planets in the diameter of Earth.

In the second stage, the surface temperature on many planets dropped very rapidly, depending on their surface area and the thermal inertia. We believe that this rapid cooling may have created an intense heat inside the core of the planets as the cores felt the enormous compressive force. This sudden compression may have resulted in the fusion processes on the planets, similar to the thermonuclear fusion of hydrogen into helium occurring in the sun at the present time. It is anticipated that the high temperature thermonuclear burning of hydrogen into helium will stop as the temperature in sun drops. On the planets, more heavy elements would have formed because of the compressive fusion process. Therefore, we are speculating that the planets' ore contain elements and compounds that contain constituents, such as oxygen, nitrogen, carbon, iron, copper, and several minerals of the heavy metal category, as a result of the compression fusion process. Many elements on the periodic table came into existence because of the extreme compression effects caused by rapid rate of cooling. With some good fortune, when hot hydrogen atoms were in the proximity of oxygen and with the correct temperature, water molecules were formed. As we shall see in stage three, water played a very important role in developing the organic compounds that lead to the development of life in cells. Among all the planets of the solar system, it happened that the Earth had many suitable conditions that allowed the progression of events to form a life support system on the Earth. Next, we shall state a few of the many conditions that favored the germination of the life on Earth.

First of all, the Earth was able to prevent the escape of gases, hydrogen, nitrogen, oxygen, carbon dioxide and water vapor, from its atmosphere because of balancing the escape the speed of gas atoms and the force of gravity from the Earth. Secondly, it had axial rotation in addition to the orbital motion surrounding the sun. The axial rotation gave day light to the parts of the Earth facing the sun and night time to the other side. Thirdly, the orbital rotation provided the seasonal changes for the weather of winter, summer and the rainy season in various regions of the Earth. Because of the intense heat, the heavy metal mass at the core of the Earth was transformed into molten mass. The mixture of the molten metal mass and non-metallic elements in the crust of the Earth resulted in the formation of lava. In the areas where the cooled surface of the Earth's layers was thin, hot lava would frequently come out of the core to the surface by volcanic eruption. This lava would cool down rapidly on the cold surface. A variety of soils, sand, and clay (black and red) may have formed on the surface of the Earth in many regions, depending on the rate at which the lava flowed and the heat was lost from the locality. We believe that the oceans and land areas were formed because of the movement of the Earth's crust and the volcanic activity over millions of years. Further, the volcanoes may have created the mountains and valleys. This is how some of the peaks and valleys of many famous mountains came into existence. The birth of mountains resulted in rain falls in land areas much farther away from the oceans and lead to the creation of the rain forest. As we explain in the next paragraph, the third stage was responsible for the development of the plant and animal life on Earth.

Now, let us discuss the causes for the origination of the life on Earth, the third stage in which it came into existence. The origin of plant life stems from the contact of organic compounds with stationary water. It is believed that a very elementary form of plant life was created inside the ocean beds initially, probably hundreds of millions of years ago. Even at the present time, the matter of great research is the study of plant and animal life at the bottom of the deepest seas. This study will potentially resolve many mysteries of life on Earth. Thus, we believe that sea weeds were the first form of life that appeared on the Earth. When the ocean beds with vegetation were exposed to the surface because of volcanic activity, the plant life, sea weeds, had the opportunity to migrate to land. Thus, many higher forms of plant and tree species came into existence as a result of the natural progression of events, the cycles of summer, winter, and rainy seasons in various regions of the Earth. The areas of the Earth with very heavy rain falls were concentrated with dense vegetation and plant life. Therefore, many types of jungles and forests, such as the tropical forests in Indonesia and the jungles of the Amazon River came into existence.

Now, let us discuss the origin of matter that transformed non-living matter into living cells. The roots of animal life come from the development of the most elementary type of life form, amoebae. The unique qualities of life cells from dead matter are life cells that can reproduce (multiply) themselves and are self directive organisms. They could consume a resource and excrete the waste. They could move randomly or decisively. Again, we believe that amoebae cells were first originated inside the water as they were considered the most rudimentary form of life. Subsequently, higher forms of life, such as insects and bacteria, were born in the proximity of the sea weeds, a basic plant life form. The cross-breeding of different insects may have developed animals, such as the fishes in the sea. The fish species evolved into even larger fish species, such as sharks. So far so good, we have answered and explained many intricate questions. We do not know how the mammal species (vertebrates) were born from the fish species (invertebrates). For instance, whales are a family of fish, but they have mammal characteristics in which a female member provides nutrition from her own body to the newly born family member. From the shallow regions of the water, amphibians, such as frogs, and big animals, such as crocodiles and alligators, came into existence in the life cycle evolution. Then, several types of vertebrate's, like cows, and thousands of other species evolved. Gradually, over centuries, the human race was originated from their ancestors, monkeys and chimpanzees. Even though the topic we have discussed has very little bearing on physics, we decided to delve into it as it is of great interest to you. Next, we shall see if it makes sense to believe the possibility that a similar human life exists on other planets in the universe.

It is evident that there is a finite, non-zero probability that life may have evolved in another star-planet system elsewhere. The reason we state this is because the steps that are occurring during the life cycle of many main sequence stars, such as the sun, are identical. Another question of interest would be whether or not other human civilizations are more advanced in their intelligence? The answer would most likely be that they would be trailing, but one cannot say for sure. Our answer is, if they are more intelligent, they would have communicated with us in a manner that we could recognize. Since in this section, we focused on the origin of the universe and the creation of the life on Earth, naturally the next topic should be the ways the age of the universe and the distances of huge celestial objects in the universe are specified and measured. Therefore, in the next section, we shall discuss the ways to improve the accuracy of Hubble's constant which is the most critical parameter in the determination of the age and size of the universe.

8.3 Deterministic Case for Hubble Constant

In the mid-1920s, a young scientist named Edwin Hubble made a remarkable discovery. While he joined the Mount Wilson Observatory in Pasadena, California as a staff member, he looked at a historic photograph of the Andromeda "Nebula". The objects that he thought were novas, he soon realized were Cepheid variable stars. By knowing the apparent brightness and luminosity of the Cepheid variable stars, astronomers can use the inverse square law to determine the distance of the stars. Based on a close examination of the photographs of several Cepheid variable stars, he declared that the universe is comprised of literally billions of galaxies. Subsequently, he categorized galaxies into four broad classes based on their appearance. Hubble summarized his classification scheme in a tuning fork diagram [17].

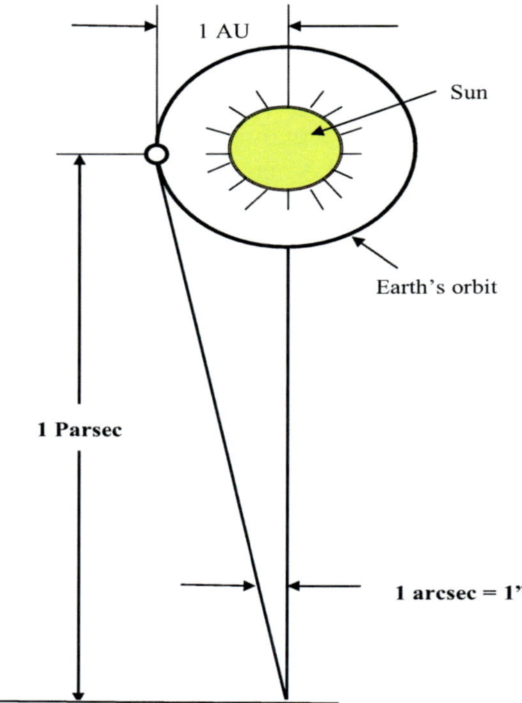

Figure 8.2 A Parsec: Largest unit of distance measurement commonly used by astronomers. The parsec is defined as the perpendicular distance at which an observer's line of sight located at **1AU** subtends an angle of **1 arc s**. (Courtesy of R. Freedman and W. Kaufmann III, Universe- Sixth Edition)

The subject matter of this section is how far these galaxies are and with what accuracy one can predict the distances of remote galaxies. Generally speaking, the distances of galaxies are expressed in the units of Parsec, the largest units of distance measure. The parsec is defined as the perpendicular distance at which an observer's line of sight, located at **1AU**, subtends an angle of **1 arc s**, as illustrated in Figure 8.2. The unit of length **1AU** is called an astronomical unit which is equal to the average distance between the Earth and the sun. This unit is used to describe the distances within the solar system.

$$\mathbf{1\ AU = 1.496 \times 10^{8}\ km = 92.96M}\ \text{miles}$$

For specifying the distances of stars, the unit light year (**ly**) is used which corresponds to the distance travelled by light in one year. This is a useful unit because the absolute speed of light (Chapter 2, Section 2.1) in space has a constant value of $\mathbf{3 \times 10^{5}}$ **km per s**:

$$\mathbf{1\ ly = 9.46 \times 10^{12}\ km = 63{,}420\ AU}$$

From the simple geometry rules, as in Figure 8.2:

1 pc = 3.09 × 10^{13} km = 3.26 ly

The astronomer employs several techniques to determine the distances of the objects of the universe outside the solar system. Because the techniques evolved out of necessity, their range of applicability overlaps. Hence, the current technique can be used to calibrate the technique of the previous generation. In Table 8.2, the distance measuring techniques and the range are summarized for different stellar objects. To study the characteristics of the stellar objects, astronomers record their spectrum. Information from the spectrum permits an astronomer to determine the brightness, distances, velocity of recession, and other variables of interest for the object.

In the 1920s, Hubble and Humason discovered that most galaxies show a red shift in their spectrum. They also found that the more distant the galaxy is, the greater its red shift and the more rapidly it is receding from our galaxy, the Milky Way. In 1929, Hubble declared the result of his discovery into Hubble's law that states:

$\mathbf{v = H_0 \times d}$ where **v** = recessional velocity of a galaxy **km** per sec
$\mathbf{H_0}$ = Hubble Constant in **km** per **s** per **Mpc**
d = distance to the galaxy in **Mpc**

We stated Hubble's law again because of its importance in astronomical studies. The law is emphasized to the greatest extent and NASA honored Hubble by naming the first orbital telescope laboratory after him, known as the Hubble Space Telescope (HST). The Hubble Space Telescope is the only telescope that is capable of producing images of astronomical objects from a wide range of signals, radio frequency, infrared and visible light wave signals received from astral objects. After the HST was launched in space, a defect of the spherical aberration in the lens of the telescope significantly reduced the signal gathering power. As a result of the defect, images of some of the stellar objects were blurred. To correct this defect, NASA sent astronauts on a space mission and fixed the problem by adding supplemental lenses to the primary objective lenses of the HST. The main advantage of Hubble's law is that it can be used as a distance indicator for remote galaxies. His law can be combined with red shift and Doppler's effect to compute distance.

The red shift for a receding object:

$$\mathbf{z = (\lambda - \lambda_0)/ \lambda_0 = \Delta \lambda / \lambda_0} \quad \mathbf{(8.5)}$$

where **z** is the red shift of an object

λ_0 un-shifted wavelength of the shifted line
λ wavelength of the spectral line actually observed from the object

Table 8.2 Distance measuring techniques and range.

Technique	Measuring Range	Objects
Stellar Parallax	2 pc-500 Parsec	Alpha Centauri-Pleiades
Spectroscopic Parallax	40 pc-10000(10k pc)	
Cepheid variables	1000 pc-30000 kpc	Center of Milky way
RR Lyrae variables	5 kpc - 100 kpc	Large Magellanic cloud
Tully Fisher Fund. Plane	700 kpc – 30000 kpc	Virgo & Comma Cluster
Type 1 Supernovae	1000 kpc – 1000 Mpc	M31, Virgo & Comma

According to Doppler's red shift formula:

$$\mathbf{z = v/c} \tag{8.6}$$

By combining this with Hubble's law, one can compute the distance of the galaxy:

$$\mathbf{d = z \times c / H_0} \tag{8.7}$$

On the basis of his discovery, Hubble presumed and declared that our universe is expanding. We agree with his law that the recessional velocity of the remote galaxy is proportional to the distance from the Earth. We disagree on the conclusion that our universe is expanding. The analogy that the distance between the dots of coins painted on the surface of a balloon increases as it is inflated by injecting air and the distances among galaxies increases with the expansion of the universe is not correct. It is only approximately true that the coin dot does not expand because the area of the coin painted on the fabric of the balloon expands with the inflation of the balloon. We think the distance between the galaxies increases in proportion to their velocity because they follow the energy conservation laws. The gravitational potential of the energy state of all galaxies is balanced by kinetic energy. Our analysis allows us to formulate a simple theorem about the state of every celestial object in the universe.

Theorem: Stability in the Infinite Universe

Our universe is comprised of a very dilute (widely spaced apart celestial objects) system of stars, galaxies, nebulae and supernovas which are in a constant state of motion. To balance the potential and kinetic energy configuration, none of the objects can stand still at any time. Also, no object may occupy the exact same location (the same space coordinates) at two different times.

Proof of our theorem is not too difficult. An amateur astrophysicist knows that the planets of the solar system orbit the sun. The sun orbits the center of the galaxy, the Milky Way, on an ecliptic with a period of 220M years [17]. It is conceivable that the galaxy, the Milky Way, will be moving and rotating in relation to other galaxies or black holes. One cannot determine the orbital period of the galactic motions because distances of separation among the galaxies is of the order of several light years. However, it is observed that the galaxies are receding and occasionally approaching, based on the measured Doppler's shift of the wavelength of light and the radiation arriving from those remote galaxies. It is observed that nearby galaxies, such as M31, are actually approaching and have blue shift. As soon as one says that the universe is expanding, it implies that the volume of space is created within the universe. One cannot explain what process creates the extra volume. We believe that it is safe to say that the remote galaxies are moving away from the local galaxies. Since no one knows the boundary of the universe, our universe is more likely to be an open, infinite universe. We see some discrepancy in Hubble's law. It was originally proposed to characterize the velocity of the recession among galaxies. We agree on the recession aspect of the law. We disagree that the recession occurs because the universe is expanding. We stipulate that galaxies are receding in order to balance the energy configurations and they are in a continuous state of motion. Our beliefs are substantiated because the uncertainty in the predicted value of Hubble's constant (**H_0 = 60-90 km/s/Mpc**) is as high as 30%. Further, the spread in the value of **H_0** is found to depend on the type of celestial object that was used to measure red shifts. For instance, the spread value is **H_0 = 40-65 km/s/Mpc** for type 1a supernovae. The measured value of **H_0** by the HST is **72 km/s/Mpc** with an uncertainty of about 10% [17] which is considered satisfactory at the present time. Even though Hubble's law brought a lot of optimism in astronomical studies, there is a lot more research needed to make further advances. In the following section, we shall discuss the limitations of the current state of technology for travelling into space.

8.4 Observable Universe: Past, Present and Future

The largest unit of measure for distances is a light year. Since no signal can travel above the speed of wave energy, we cannot measure the changes that occur at a rate that is faster than the speed of light. The inability of science to detect changes above the speed of light, places severe limits on the rate at which space vehicles can travel. At high speeds, the space vehicles' speed and direction should be maneuvered in a very short time. It is imperative that the control rockets receive advance warnings of intervention in the projectile trajectory of spacecrafts. For real time control considerations, when maps of the universe are created, they should be partitioned into three segments.

1. Present universe –the size of the universe with object distances one light second away
2. Past (proximity) universe – the size of the universe distances one light day away
3. Future (intergalactic distant) universe – the size of the universe distances more than a light day away.

Those objects that are within a distance of one light second (coincidently, it is the approximate distance of the moon from the Earth) may be observed without perturbation in their physical position in real time. For those celestial objects that are one light day away (**3600 × 24 × 300,000 = 25920M km**), controlling the system may be achieved based on a short history of events. For those celestial objects that are more than a light day away, it is not possible to send control signals in real time to exercise control. We classified this universe as the future universe because the technology of the future will be required to make any progress. Actually, at the present time, the celestial objects within the future universe can be observed only from their distant past. To illustrate this fact, let us consider the case of the summer triangle. Astronomers routinely observe the position of a trio of stars, most popularly known as the summer triangle, that are employed to identify the position of several other celestial objects. The summer triangle is comprised of three stars, Vega, Altair, and Deneb. This trio is predominantly visible in the northern hemisphere in a clear night sky. Based on luminosity, the distances of these stars are 16.8 light years for Altair, 25.3 light years for Vega, and 3230 light years for Deneb. Because the information received from these stars is skewed in time to such an extent, the real instantaneous position of any of these stars is impossible to determine. This example proves that travelling to even our next neighboring star is an incredible challenge, even if one could possibly commute at the speed of light. In the next section, we shall discuss a very important topic for the human race, the fate of the solar system and how long civilization on the Earth shall survive.

8.5 Fate of the Solar System and Our Civilization

As stated before, our solar system and the sun were born approximately 4.5B years ago. It is our assertion that a smaller star collided with a big star, resulting in our sun and the planet system. For hundreds of years, scientists have been trying to determine when our solar system, the sun, and our civilization will cease to exist. The present estimate, from Chandrasekhar and others, is approximately after another 4.5B years when our sun will become a cold white dwarf star [26]. Also, scientists are unable to answer if the mass of the sun is depleting as a result of the loss of billions of watts of radiation energy into space. Further, it was not clear which process sustains the loss of radiation energy from the sun. Einstein's mass energy conversion equation inspired astronomers to answer the question that they thought was brilliant. Astronomers now believe that at very high temperature, **15M°K,** hydrogen atoms in an ionized state are fused together to form helium atoms at the core of the sun. This fusion process is known as the thermonuclear burning of hydrogen into helium. This is described as a three step process [17]. According to the description of the reaction, **0.7%** of mass of the hydrogen atom is converted into radiation energy. Astronomers explained the vast amount of energy released from the sun by applying Einstein's equation, $\mathbf{E = M \times c^2}$. As per this equation, the small mass is transformed into a huge amount of energy. Still, **0.7%** of mass conversion for each formation of a helium atom is a very high fraction. What puzzled us was that if the mass of the sun is changing because of the loss of energy by radiation, why has the Earth's period of rotation surrounding the sun not been affected for the past several centuries? Therefore, the thermonuclear fusion and burning of hydrogen into helium atoms and, the mass lost from the conversion of electrons into energy is not a satisfactory explanation for the luminosity of the sun.

The answer comes from the third postulate of special **Skylativity®**. We claim that whenever a radiation event occurs, it happens because the electrons of gas atoms in a high potential energy state make a transition to a low potential energy state. There is no conversion of the mass into energy during the emission of radiant energy.

Theorem: Mass Conservation in the Universe

The matter inside white stars, such as the sun, can neither be created nor can it be destroyed in spite of the loss of billions of watts of radiation energy per second. The mass of the white star, the sun, always remains constant unless a portion of its mass is dislocated by another star on a collision course. The self force of gravity of the sun prevents the leak of atoms into space (vacuum) though the surface temperature of the sun is as high as **5800°K**.

Proof: The basis for proof of this theorem lies in the roots of Newton's law of gravitation and Newton's form of Johannes Kepler's third law. According to Newton, the force of gravity among two objects is proportional to the product of their masses and inversely proportional to the square of the distance of the centers of gravity of both objects. The proportionality constant **G** was known as a universal constant because the law holds in the universe and it holds for every object of any size in the universe, except for energy waves whose rest mass is zero. Semantically:

$$\mathbf{F_g = G\,(M1 \times M2)/\,R^2} \qquad \mathbf{(8.8)}$$

It is a well known fact that the measured solar year period, or the time required for the Earth to complete one rotation surrounding the sun, is observed to be constant for the past hundreds of centuries. Both the tropical year and the sidereal year periods have not deviated from their values by the smallest fraction of a second over centuries. From Newton's form of Kepler's third law, which relates the sidereal period and the masses of rotating objects, it is clear that if a period is constant it implies that the mass of the objects should be constant if the semi-major axis is constant. The formula for the sidereal period for the Earth's orbit surrounding the sun from Kepler's law is as follows:

$$\mathbf{P^2 = [4\pi^2/G(M1+M2)] \times a^3} \qquad \mathbf{(8.9)}$$

where **P** = the sidereal period of orbits in seconds
a = the semi-major axis in **m**
M1 = the mass of the first object in **kg**
M2 = the mass of the second object in **kg**
G = the universal constant of gravitation = $\mathbf{6.67 \times 10^{-11}}\ \mathrm{m^3\,kg^{-1}s^{-2}}$

To prove our theorem we must prove that the sidereal period of the Earth's orbit surrounding the sun does not alter over a period of many decades. If the measured value of the sidereal period, year after year, does not change (increase) by **100ps** per year, that would imply, from the previous two equations, that the mass of the sun is not lost according to Einstein's STR prediction. How we have arrived at the magic number of **100ps** will be evident shortly.

In the following, we shall formally prove and validate our theorem by comparing the computation of the changes in the sidereal period of the Earth's orbit to the measured experimental value. To accomplish our objective, first we shall compute the total mass that is lost per second from the sun in the thermonuclear burning of hydrogen into helium that will generate power on the sun to achieve measured luminosity. We shall utilize the equation **$E = m \times c^2$** to compute the loss of mass equivalent of energy from the sun because of the mass to energy conversion. It turns out that in fusing hydrogen into helium, six hydrogen isotopes need to be fused together to form one helium atom, and two hydrogen atoms along with γ ray energy is released as a result of annihilation of two electrons and two positrons. One fact that is obscured about the thermonuclear burning of hydrogen into helium, is when two (light weight) hydrogen isotopes are fused together, the mass of the combined one **^2H** isotope, one proton and one neutron (neutron mass = **1.67493×10^{-27} kg**), is lower than the mass of **2 ^1H** hydrogen isotopes (protons mass = **1.67262×10^{-27} kg**) that started the reaction, plus the fusion releases a positron. This does not make sense (refer to Figure 8.3).

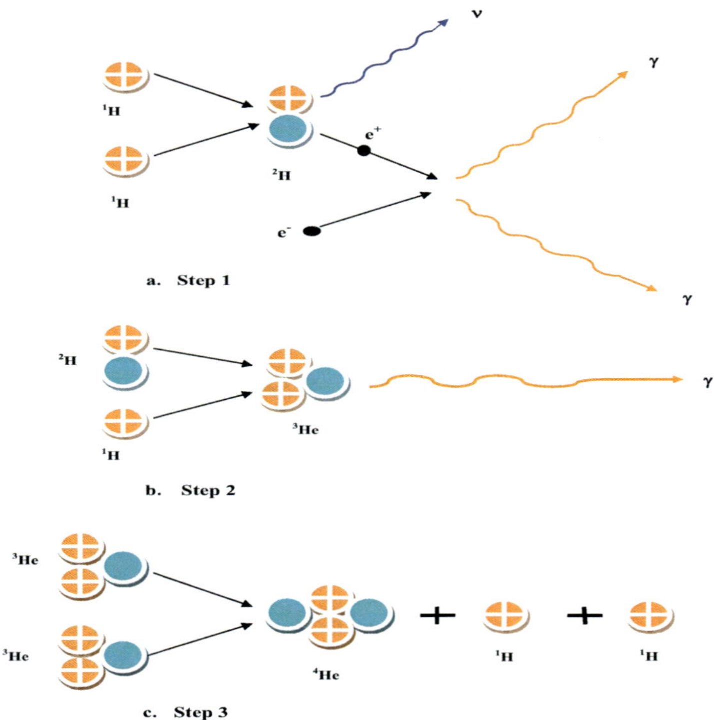

Figure 8.3 Thermonuclear fusion of hydrogen into helium, 3 energy release steps. (a) Two protons fuse to form an isotope of hydrogen. (b) Two protons fuse to form an isotope of hydrogen. (c) Two ^{3}He nuclei collide to form the ^{4}He isotope and two protons.

Therefore, further analysis of the events occurring during the thermonuclear fusion is required. The examination of current events does not clearly identify the balance of proton and neutron masses down to the quark level. The final composition of protons and neutrons is comprised of quarks according to the extension of the standard model. In the event, when a proton is transformed into a neutron, a negatively charged electron mass is added to the proton to neutralize the proton charge which would result in a gain in mass for the proton producing the neutron that is heavier than the proton. Otherwise, the positron from the proton should be detached from the proton that should decrease the mass of the proton and the end product neutron. Further, according to the quark extension to standard model, protons are comprised of two up quarks and one down quark. The neutrons are comprised of two down quarks and one up quark. According to recent measurements, the mass of the up quark is (**0.003 GeV/c^2**) and the mass of down quark is (**0.006 GeV/c^2**) which is two times the mass of the up quark. If one adds up the number of masses it does not explain why the total mass of the helium atom isotope, after the fusion reaction, is lower than the total mass of hydrogen atoms before the reaction, even though the helium isotopes contain two neutrons. It is not obvious why the neutrons bound in the helium atoms, with the same quark configuration as an individual neutron, have lower mass than an unbound neutron. It is surprising that the mass numbers for quarks on both sides of the thermonuclear burning equation would balance if mass numbers of both up and down quarks is equal to **0.003 GeV/c2**. Figure 8.4 illustrates the structures within an atom down to the most fundamental particles, quarks. The distances among the particles are not to scale because of the enormous differences in the sizes. Also, the approximate mass and charges for the various quarks are included in Table 8.3.

Table 8.3 Approximate mass and charges on quarks.

Quarks Flavor	Approximate mass GeV/c^2	Electric Charge q
u up	0.003	2/3
d-down	0.006	-1/3
C charm	1.3	2/3
S Strange	0.1	-1/3
t top	175	2/3
b bottom	4.3	-1/3

The second confusing fact that is not answered by present scientific theory is why all of the positrons created during the thermonuclear fusion process are immediately annihilated with electrons and none could escape from the sun. Even the tiniest neutrino particles of some Eigen states have escaped from the sun, in spite of their very low mass and almost no reactivity, and are detected on the Earth. For that matter, it is surprising that all of the types of anti-particles are not found naturally. It should be of prime interest to quantum physicists to further analyze the wavefunction of antiparticles during the thermonuclear fusion process and to study the escape probability for positrons from the sun by using Schrödinger 's equation. This study may provide very useful information at the interior of the sun.

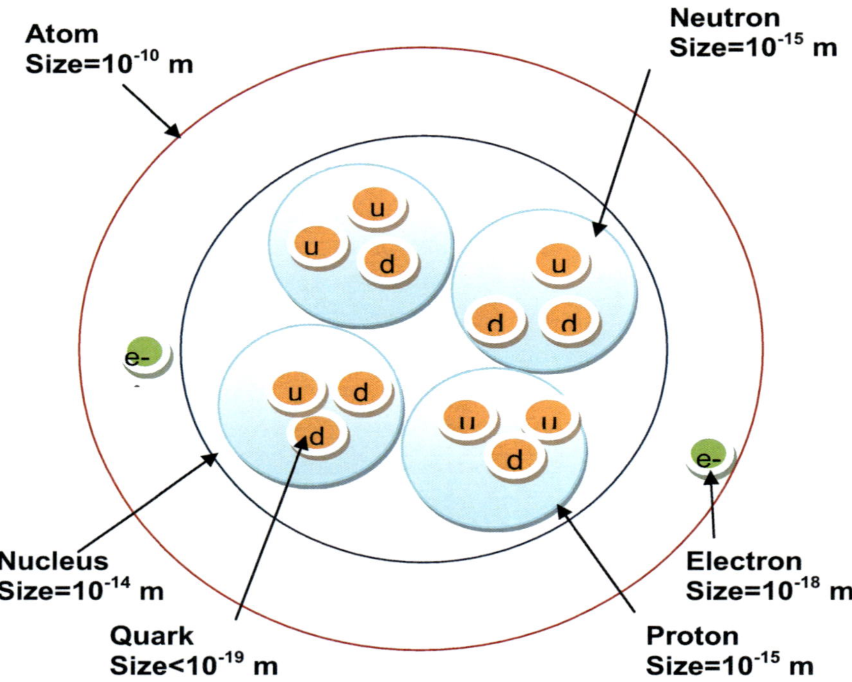

Figure 8.4 Structure within an atom down to fundamental particles quarks. (Adapted from a chart of the standard model for particles and interactions. http://CPEPweb.org)

For the time being, without going into detailed analysis, we will accept the bottom line, the fact that the mass of two electrons and two positrons is lost and converted into energy in the production of one helium atom. From box 18-1 pg. 393 [17]:

mass of four hydrogen atoms = $\mathbf{6.693 \times 10^{-27}}$ **kg,** subtracting
mass of one helium atom = $\mathbf{6.645 \times 10^{-27}}$ **kg**
mass lost = $\mathbf{0.048 \times 10^{-27}}$ **kg**

this lost mass corresponds to energy $= \mathbf{m \times c^2}$
$= \mathbf{0.048 \times 10^{-27} (3 \times 10^8)^2}$
$= \mathbf{4.3 \times 10^{-12}}$ **joules**

This represents energy produced from the conversion of **0.7%** of the mass of hydrogen into energy. Therefore, when **1 kg** of hydrogen is consumed to form helium, **0.007 kg** will be converted into radiation energy and the
Corresponding energy = **0.007 kg** $\times$ **(3** $\times$ **10^8)^2** = **6.3** $\times$ **10^{14} joules**, will be produced. To give the current luminosity to the sun, **3.9** $\times$ **10^{26} joules** per second, power should be produced by burning hydrogen. The power emitted from the surface of the sun is divided into 40% visible light, 50% IR, 9% UV, 1% X-ray, radio, etc. [30]. This allows us to compute the loss of mass on the sun per second which is given by

$(3.9 \times 10^{26})/(6.3 \times 10^{14}) = 6 \times 10^{11}$ **kg** per second.

Our objective is to find out the deviation of the sidereal period because of the change in the mass of the sun per year.

From the parameters:

mass consumed per year = mass consumed per second $\times$ number of seconds per year
$= 6 \times 10^{11} \times 3.16 \times 10^7$ **kg**
$= 18.96 \times 10^{18}$ **kg**

This number appears to be very small in comparison to the mass of the sun that is
1.989 $\times$ **10^{30} kg**, the number accumulates over the life of the sun. It is estimated that the current age of the sun is 4.5B years.

The loss of mass of the sun at the present time is:

$$18.96 \times 10^{18} \times 4.5 \times 10^9 \text{ kg} = 85.32 \times 10^{27} \text{ kg} = 0.853 \times 10^{29} \text{ kg}$$

which is 4.289% of the mass of the sun.

Now let us calculate the impact of the loss of mass of the sun on the sidereal period of the Earth since the beginning of the solar system and the difference in the sidereal period each year. We shall compute the period by substituting numbers in Kepler's third law formulae:

$$P^2 = [4\pi^2/G(M1+M2)] \times a^3$$

Substituting values of parameters we get

$$P^2 = [4 \times 9.8696/6.6726 \times 10^{-11}(5.974 \times 10^{24} + 1.989 \times 10^{30})]\ (1.521 \times 10^{11})^3$$
$$= 39.48 \times 3.519 \times 10^{33}/(6.673 \times 5.974 \times 10^{13} + 6.673 \times 1.989 \times 10^{19})$$
$$= 138.93 \times 10^{33} / (39.86 \times 10^{13} + 13.273 \times 10^{19})$$

Since the ratio of the solar mass to the Earth mass is **0.333** $\times$ **10^6**, we can remove the first term in the denominator to simplify the arithmetic:

$\mathbf{P^2}$ $= \mathbf{138.93 \times 10^{33} /(13.27259 \times 10^{19})}$ giving
$\mathbf{P}$ $= \mathbf{(10.467 \times 10^{14})^{1/2} = 3.23534 \times 10^7}$ Seconds

without removing the first term:
$\mathbf{P^2}$ $= \mathbf{138.93 \times 10^{33} /(13.27263 \times 10^{19})}$

which gives us:
$\mathbf{P}$ $= \mathbf{3.23489 \times 10^7 s}$

The ratio of these two values is **0.99986**. Therefore, removing the first term should provide meaningful results for comparison purposes. It is noteworthy to see that the exact number of seconds in the sidereal period is **31557600**. Therefore, in terms of the absolute values, Kepler's third law provides period values that are only approximately correct. Let us look at the impact of the loss of the mass of the sun on the sidereal period for the decrease in mass to date. Since the mass decrease to date is 4.289%, the new value for the sidereal period is:

$$\mathbf{P^2} = \mathbf{(138.93/(0.95711 \times 13.27259)) \times 10^{14}}$$
$$= \mathbf{10.93650 \times 10^{14}}$$

which gives us:
$\mathbf{P}$ $= \mathbf{3.307038 \times 10^7 s}$

The results of our computations indicate that the sidereal period should have increased 4.7937% over a period of 4.5B years! The difference is approximately **1512779.49s** or 1.**5Ms** since the beginning of time. This translates to **333 μ-s** increase per year. We cannot see the experimental data that far back in time, we shall compute the sidereal period change, year after year. Also, to verify the **333 μ-s** numbers, let us calculate the difference in the sidereal period for two successive years. Performing computations for the sidereal period, year after year, is not a trivial task because the decrease in the mass of the sun per year is very small compared to the entire mass of the sun, to perform arithmetic with acceptable accuracy. However, very accurate cesium **Cs** clocks have been designed that can measure the number of seconds per year, every year, with accuracies more than a few nanoseconds. Therefore, we suggest that an experiment should be performed to determine the sidereal period each year.

A national standard institute, such as NIST, that maintains these clocks, should measure the time difference, year after year. Our prediction is that time will not change by **333 μ-sec**, as predicted by Einstein. We believe that there is no loss in the mass of the sun and hence, the length of the year should be constant over millions of years. The experimental results of the sidereal period measurement should validate our work if the difference in the sidereal period for each successive year is much lower, for instance, below **1μ-s**. More recent data from the experimental measurement of the sidereal period indicates that the sidereal period indeed increases by **0.4ms** (**400 μ-sec**) each year because the mass of the sun is diminishing. This results in a tiny increase of the distance of the Earth from the sun. The astrophysicists believe that this small increase in the sidereal period is of no consequence as it is very small and does not have any other effect. We wonder if this increase of the distance of the Earth's orbit should decrease the temperature on the surface of our planet and should have a global cooling effect.

Now, let us calculate the life span of the solar system, based on the computation from the fraction of hydrogen mass that is available on the sun for conversion into **He** and energy, and the mass consumed per year to sustain the present level of luminosity. Of course, our calculations are based on the assumption that the minimum temperature required for the thermonuclear fusion process is **1M°K**.

From the computation in this section, the mass consumed per year to generate the required power at the sun == **18.96 × 10^{18} kg**.

From Section 2.7:

The mass inside the sun that is available for conversion
= **0.8 × solar mass × 0.07 kg = 0.56 × 1.9891 × 10^{30} kg**

Life span== **1.113896 × 10^{30}/(18.96 × 10^{18})**
== **58.749789 × 10^{9}** years
== **58.749789B** years

If we assume that the minimum temperature required for the thermonuclear fusion of hydrogen into helium is **10M°K**, the new value for the remaining life span is:

Life span== **(0.3 × 1.9891 × 10^{30} × 0.07) /(18.96 × 10^{18})**
== **2.203117089B** years

What we have shown is that the life span computations critically depend on the minimum temperature at which the thermonuclear fusion reaction occurs. The life span changes by a factor of **30** if the temperature requirement of the fusion changes by a factor of **10** from **1M°K** to **10M°K**. Realistically, the minimum temperature for the fusion reaction to occur is approximately **5M°K**. The corresponding available mass fraction of the hydrogen gas for the conversion is **0.7**. This gives a life span value of:

Life span== **(0.7 × 1.9891 × 10^{30} × 0.07) /(18.96 × 10^{18})**
== **5.140606B** years

This value agrees with the data published for the life span of the solar system in widely published literature.

For more precise computations, the minimum temperature at which the thermonuclear burning of hydrogen into helium takes place, one must compute the probability of the fusion at a specific pressure. This probability depends on the parameters, such as the mean free path and the velocity of the hydrogen atoms in a plasma state in which the atoms of hydrogen are completely ionized. The fusion probability number may be utilized to calculate the minimum energy required for the fusion to occur. Alternately, one can use the mean free path (MFP) distance to calculate the potential energy between two ionized hydrogen atoms (page 1337) [22]:

$$\textbf{MFP U} = k_e \times q_1 \times q_2/r \qquad \textbf{(8.10)}$$

where k_e Coulomb's constant = $\mathbf{8.99 \times 10^9\ N\ m^2/C^2}$
$\mathbf{q}$ Proton charge = $\mathbf{1.6021 \times 10^{-19}\ C}$
$\mathbf{r}$ separation distance to overcome the repulsive Coulomb force for hydrogen ions

For the collision to occur, this energy must be equated to:

Thermal energy == $\mathbf{(3/2) \times k_b \times T}$

From equation (8.10) one could solve the minimum temperature **T**. Perhaps more accurate analysis is required to increase accuracy of computations to derive value of minimum temperature at which thermonuclear fusion of hydrogen into helium is occurring.

So, what is the fate of the solar system and our civilization? As per the prediction from our theorem, and by Einstein's computation of the mass to energy conversion, the mass of the sun does not change significantly. This is the case even though a tremendous amount of wave energy is radiated for billions of years. Further, it is highly probable that the sun will continue to provide the radiation energy by processes similar to hydrogen burning, such as helium burning, carbon burning, and oxygen burning, in the post hydrogen to helium conversion era. It is likely that these other processes may occur at a lower temperature than the thermonuclear burning of hydrogen into helium.

The good feature of the thermonuclear burning of hydrogen into helium process is that it creates two ions of hydrogen atoms at the end of the process. This results in a chain reaction for the thermonuclear fusion process. The chain reaction allows sustaining the process as long as the hydrogen atom supply is available. Ultimately, the thermal energy of the sun will be reduced below the level that would not support any kind of thermonuclear reaction. Therefore, the core temperature of the sun will cool off. Gradually, all of the hydrogen will be consumed and transformed into helium and heavier elements. We believe that the core of the sun will slowly be transformed into a liquid phase. Eventually, it will solidify and will become a cold white dwarf star.

When this happens, our solar system will become cold and the Earth beings will freeze to death. As per the current estimate, this will happen after another 4.5B years, or possibly longer, if the other burning processes continue. This estimate may not be very accurate. To predict the life span of our sun with higher accuracy, it is very important to determine the minimum temperature at which the thermonuclear burning of hydrogen into helium atoms may occur. The lower limit on the temperature at which the thermonuclear burning of hydrogen into helium gas occurs, will determine the mass fraction of the total hydrogen mass on the sun that is available for the conversion process. The task of determining the minimum temperature at which the hydrogen to helium conversion may occur by the fusion process is difficult. From this temperature, and after looking at the best model for the temperature profile for the core of the sun, one could compute the amount of hydrogen that is available for conversion into helium at the temperature required for the process to occur. Therefore, in the mean time, it is imperative that new technologies should be invented to create the source of thermal energy which will keep the Earth warm. Further, bioscience should invent process alternatives so that trees and leaves transform carbon dioxide into oxygen via artificial photosynthesis.

Now that we know that the fate of our solar system crucially and ultimately depends on the fate of the sun, let us look at the predictions of modern theories about what will happen next. The modern theories suggest that the white dwarf stars are the final state for most stars after they exit from the main sequence star state. The nuclear fusion process of turning hydrogen into helium on the main sequence star, such as the sun, ends due to the lowering temperature or the lack of hydrogen fuel, and the star becomes red giant star. During the main sequence state, stars spend most of their existence in a state of equilibrium. The two forces, the gravitational force trying to pull inward and the thermal pressure from the nuclear reaction at the core trying to push outward, balance each other. When the nuclear fuel within the star is exhausted, it must contract. The end product for this collapsing star is a low mass white dwarf star. Before the contracted star becomes a white dwarf star, the star goes through a transition phase known as the red giant phase. The contraction in the core increases the temperature that causes the burning of the helium shell core into carbon and oxygen. It is believed that the mass of the white dwarf star can be no more than 1.4 times the solar mass, as stipulated by the Chandrasekhar limit. In the red giant phase, the sun will grow in size and will eject some of its mass. If a star would be more massive than the sun, up to four times the solar mass, it could lose enough mass to form a white dwarf star with mass no greater than **1.4** solar mass. The stars those are more massive than four times the solar mass will not end up being white dwarves. These stars collapse even further, becoming a neutron star or a black hole. Their final moments as "normal" stars are more spectacular. Most of them, it is believed, will finish their lives in a supernova explosion.

As explained before, the sun may blow up in size, expanded to near the Earth's orbit **(300,000 km)** before it will become a white dwarf star the size of **10000** to **20000** miles in diameter. The white dwarves are known to have a wide range of temperature, from **70000°K** to **5000°K**, although most fall between **8000°K** and **10000°K**. Also, the white dwarves have a lot smaller brightness than a star in the normal main sequence state. The luminosity of a white dwarf star from the sun will be as low as 1% of the present day brightness. Next, we shall address a slightly different topic, the direction of the rotation of the planets on their axes. It is known that the Earth, in addition to the orbital motion surrounding the sun, rotates on its axis, from the west to the east. There are some planets, such as Venus, Uranus, and Pluto, which rotate from the east to the west. In the following section, we shall analyze the reasons behind the retrograde rotation of these planets on their axes, from the east to the west.

8.6 Planets with Retrograde Axial Rotation

Before we can explain the opposite axial rotation of the planets Venus, Uranus and Pluto, let us explain why a planet like Earth rotates on its axis, in addition to the orbital rotation surrounding the sun. As per our explanation, the solar system and our planetary system were formed as a result of the collision of two stars approximately 4.5B years ago. At that time, all of the planets must have been in a gas form like our sun is now. Because the planets were so small, there was no way the thermonuclear burning of the hydrogen atom into the helium atom process would occur, a process that is happening on the sun. Therefore, the temperature of the planet's surface and interior started dropping. The planets receded from the sun, to obey the laws of gravitation and they started orbiting surrounding the sun. They were prohibited, moving further away by two balancing forces, the inward gravitational pull and the outward centrifugal force, because of their orbital speed. Essentially, their orbit radius was determined by the escape velocity as computed by the formula of ordinary mechanics.

We believe that in the beginning, the planets were not rotating on their axes. All of the planets were facing one side toward the sun. When the planets were hot, the gases on the planets were not able to escape because they were under the influence of the distributed granular self gravitational force effect of the planet balancing the velocity of gas molecules due to thermal agitation. The gravity effects were granular, as the entire planets were in a gaseous form. According to the first law of thermodynamics, the heat propagation through a medium always flows from a high temperature location to a low temperature location. Therefore, the tendency of heated gas molecules from the side of the planet facing the sun was to flow toward the dark side. Now, according to the second law of thermodynamics, the total entropy of an isolated system always increases in irreversible processes. An increase in entropy implies an increase in the random behavior or a decrease in order. Also, the motion of thermally agitated gas molecules is a natural, irreversible process. Usually, the movements of the gas molecules were random and would prevent the axial motion of the planets. However, as the temperatures of the planets were lowered, the core became a liquid and solidified because of the compressive forces induced by self gravity. Now, gas molecules on the planet's atmosphere and surface were under the influence of the localized pools of gravitational fields. These pools of gravitational field force effect and the directional heat flow commanded by the first law of thermodynamics resulted in a more uniform motion of the vast number of gas molecules. The uniform movements of the gas molecules were accumulated to create a torque. We believe that two factors, the forces of the localized coarse gravitational pull and the heat flow developed a bias for the direction of the torque. In terms of the proximity, the entropy of the system decreased even though the isolated system of the planet was experiencing an irreversible process. This is an apparent violation of the second law of thermodynamics. Strictly speaking, the system of the planet is not isolated, as the bright side of the planet facing the sun received a lot more thermal energy than the dark side. Therefore, the second law of thermodynamics is not violated when the total planet system environment is considered. Figure 8.5 illustrates the direction of rotation on the axes of all the planets in the solar system. Also, we have included the masses and the sidereal period in years relative to the mass of the Earth $\mathbf{E_m = 1}$, and the period for the orbital rotation of the

Earth $\mathbf{E_y = 1}$ as one year for every planet.

Now that we understand why the planet achieved the axial motion, in addition to the orbital rotational motion surrounding the sun, it is easier to see why some planets' axial motion is in the opposite direction, from the east to the west. Obviously, the center of gravity and the center of rotational moment of inertia were heavily affected by the mass distribution of the planets. The unbalance in the mass and the center for the rotational inertia, created a bias toward the direction of rotation, from the east to the west or the west to the east. This would explain why the planets, such as Venus, Uranus and Pluto, rotate on their axes in the opposite direction, from the east to the west, as compared to our Earth's axis of rotation, from the west to the east. This explanation makes complete sense and is logically consistent with the laws of thermodynamics. In the next section, we shall answer another interesting question about the asymmetric orbits of the comets surrounding the sun. It is well known that many comets have very oblong orbits with large eccentricity. We shall explain the reasons for these abnormal orbits of comets.

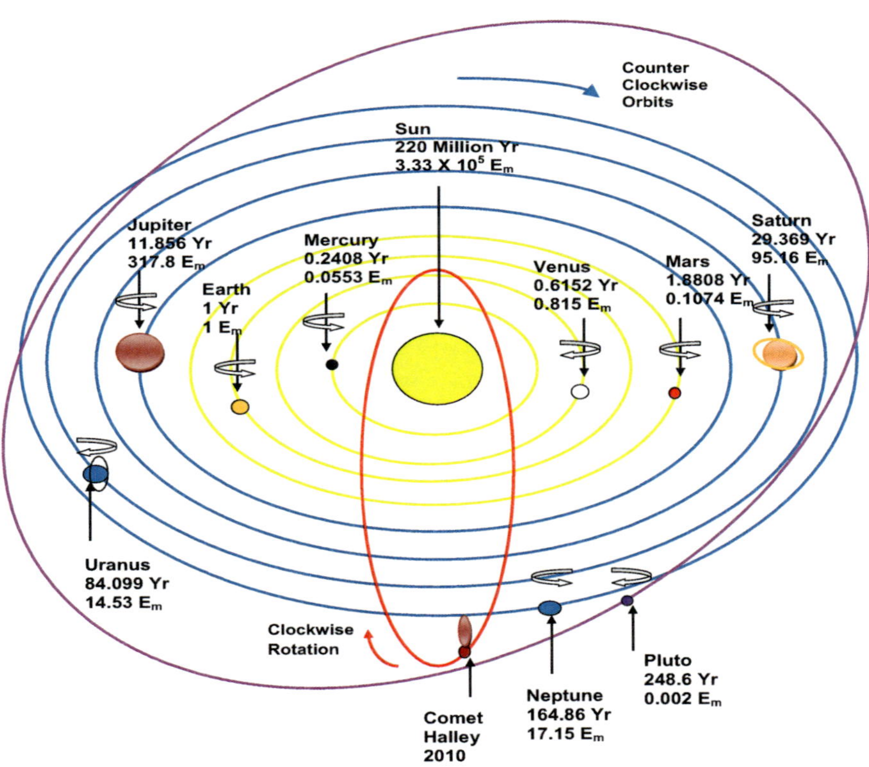

Figure 8.5 Direction of the rotation of the planets on their axes, masses, and orbital period compared. The planets Pluto, Uranus, and Venus rotate in the opposite direction on their axes, from the east to the west.

8.7 Comet's Asymmetric Orbit

Comets have been an object of astronomers' interest for their unique orbital motion since the early 18th century, when English astronomer Edmund Halley, in 1705, first spotted the famous Halley's Comet, named after him. He applied Newton's theory of the gravitational perturbation to calculate the orbital period of comets and concluded that one of the reasons for the asymmetric orbit of comets is the force of gravitation from the planets Jupiter and Saturn. Since the discovery of Halley's Comet, few other comets, Borrelly's, Shoe maker-Levy 9, and Hale-Bopp, have been found. More recently, four comets have been visited by NASA's spacecraft ICE, passed through the tail of Comet Giacobini-Zinner in 1986 and Comet Grigg Skjellerup in 1989. Another spacecraft, Deep Space 1 (DS-1) imaged the nucleus of Comet Borrelly in 2001. In 1986 five spacecrafts from the USSR, Japan, and the European Space Research Organization (ESRO) visited Halley's Comet. Of all the comets, the orbital period of Halley's Comet is regular and the average is 76 years which could vary as much as 79.3 years between the years 239 BC to 1986 AD. Even for Halley's Comet, one cannot compute the dates of reappearance by simply subtracting multiples of 76 years from 1986 for the above mentioned variations. The next appearance of a comet is anticipated to be in early 2062. In the following, we shall examine the reasons why the orbits of comets are so elongated.

One of the plausible explanations for the elongated orbits of comets is that comets have a very large tail of gases. We believe that this large tail of gaseous matter accompanying the comets shifts the center of gravity of the comet from the solid mass inside the comet to the gas tail region. When Halley's Comet passes by the large planets Jupiter and Saturn, the asymmetry of the center of mass of the comet results in a sling action and pushes the orbit of the comet far out. Also, Halley's Comet revolves its orbit in a retrograde motion of clockwise, as opposed to the Earth's counter-clockwise rotation for reasons similar to the reasons explained for Venus. We shall complete this discussion of the comets' orbit by the following statistics for Halley's Comet.

Perihelion distance:	**0.587 AU**
Orbital eccentricity:	**0.967**
Orbital inclination:	**162.24°**
Orbital period:	**76.0 years**
Next perihelion:	**2061**
Diameter:	**16 × 8 × 8 km**

In the next section, we shall discuss the purpose and limitations of the string theory, originally proposed by Dr. Brian Greene, a prominent physicist of 21th century. Modern day physicists believe that the quest for the ultimate unified field theory has been discovered in the seeds of the string theory. In essence, the formulation of the string theory has created the union of laws of the large, general theory of relativity, and the laws of the small, quantum mechanics, with an admirable level of success.

8.8 String Theory and New Dimensions

The limitations of the standard model inspired physicists to invent new theories, providing solutions to many unanswered problems. The great success of the standard model stems from the fact that by applying that model, the properties of all the elements found in nature and the elements found on the periodic table, or the compounds artificially fabricated using chemical, physical, and biophysical processes, are very well understood. The standard model satisfactorily describes all of the particle phenomena whenever the calculations of the theory's predictions and the experiments can be reliably done. Even though the standard model answers many questions, it is not able to answer two major questions. What determines the masses of the fundamental particles, quarks, leptons, $\mathbf{W}^{\pm}$, and $\mathbf{Z}^{\pm}$ bosons? And, what is the cause of gravity? Further, the standard model does not include the gravity of large objects (weak interaction forces) with the strong forces of charge and the magnetic spin of charged particles.

Many physicists are convinced that the unanswered questions could be answered if we can find the extra dimensions to characterize the effects of the attributes of objects such as the mass and the force of gravity exerted by the mass. Dr. Brian Greene developed his string theory by thinking out of box about three spatial dimensions and one time conventional dimension. His theory proposes that our universe has 11 dimensions in all, and those are curled up and if unfolded will answer all of the questions. Because of our limited visibility, we are unable to sense the additional dimensions above the three spatial dimensions and one time dimension. We do not necessarily agree with the idea of more spatial dimensions or dimensions that are hidden because of their small size. The reason is the new dimension must manifest its effect on existing visible dimensions to some degree by process of accumulation. We do agree that other dimensions exist and are not known to humans to date. An example of a new dimension would be the force of life or spiritual energy. Of course, this dimension would not answer the fundamental questions, the reasons for mass and the reasons for gravity force. Yet another unanswered question from the standard model is how fast the force of gravity propagates. The majority of modern day physicists agree that the force of gravity propagates at the same the speed as the speed of light or electromagnetic radiation. Because we have postulated that the speed of light is variable and depends on a frame of reference, it does not make sense that the speed of the propagation for gravity is the same as the speed of light. Further, the speed of the propagation of light depends on the index of refraction. There is no such property in existence for the gravitational field. Therefore, we propose that the propagation of the speed of the gravitational field is the same as the absolute or true the speed of light in a vacuum, according to the procedure specified in Section 2.1.

8.9 Summary

In this chapter our intention was to show that the prevailing concepts for the nature of black holes that possess the super-massive force of gravity do not make sense. Another topic that we addressed is the origin of the creation of the universe. As per the present belief, that our universe was created as a result of one big explosion and the subsequent expansion, this does not make sense. We believe that the light waves do not escape from the black holes because they are absorbed by dark matter. Also, we explained that because both the size and age determination of our universe are evolving facts and that something cannot be created from nothing, the big-bang theory for the creation of the universe is insensible. Throughout this book, our purpose is to provide the answers to many difficult questions that are not known to physicists by applying the solid logic and determination of facts that are consistent with the prevailing proven theories and practices.

In Section 8.1, we began by explaining the nature of the black holes with huge sizes and gravitational pull. According to prevailing understanding, they are so massive that light cannot escape from the force of gravity of a black hole. We portrayed a different view. We determined that light is not trapped, but is more likely to be absorbed. Our computation of the Schwarzschild radius (event horizon distance) for a hypothetical black hole constructed from the densest matter proves that light shall not be trapped if the radius of a spherical black hole is larger than **16.42** miles. In practice, it is discovered that the majority of black holes are of a lot larger radius than the computed radius of **16.42** miles. Also, it is understandable that the density of the black holes may not be as high as the density of a neutron. Thus, we proved that light is absorbed by all black holes and not trapped. In section 8.2, we expressed our doubts about the validity of the existing big-bang theory for the creation and origin of the universe. We believe and concluded that our universe is open and infinite, and its beginning and end cannot be determined. However, we could predict that approximately 4.5B years ago, the solar system and the Earth were formed as a result of the collision between two stars. It is highly probable that other stars, galaxies, and supernova were formed because of multiple big-bang events, similar to the collision of two stars that formed the solar system.

In Section 8.3, we addressed the issue of predicting the accuracy of Hubble's constant, which is a proportionality constant of Hubble's law. Hubble's law states that the recessional velocity of the receding galaxy is directly proportional to the distance of the galaxy from our own galaxy, the Milky Way. His law implied that our universe is expanding. We presented a different view in that the galaxies are moving away from each other at a velocity directly proportionate to the distance from the Milky Way. Our conclusion was based on the fact that there is no evidence of the expansion in distances among the stars within the Milky Way. One of the major hurdles, as it relates to space travel, is the size of the universe and the distances between the stars in our galaxy. It turns out that the nearest neighboring star to the sun, Alpha Centauri, is approximately 4.5 light years away. It means that it is as far away as the distance at which the speed of light would travel, 4.5 years. Therefore, any space mission that involves travel outside our solar system should require the speed of travel far beyond the current generation of spacecrafts. Also, the sensory systems employed on these spacecrafts should have very high the speed of response. In Section 8.4, we suggested that the maps of the sky should be partitioned into three regions, present universe, past universe, and future universe, based on their sizes. The mission involving travel to the different regions will require a different set of technology and instrumentation to control and program the mission objectives.

In Section 8.5, we concentrated on the fate of our solar system; this is an obvious question for survival of human civilization. Our conclusion was that the current estimate for the remaining life span of the sun and the life on Earth, of another 4.5B years, may be incorrect. The current estimates are based on the end of the thermonuclear burning of hydrogen into helium. When the supply of hydrogen gas on the star sun is consumed, the high energy producing conversion will stop and the entire solar system will eventually be frozen. We believe that the present estimates are optimistic because the thermonuclear burning would stop when the temperature on the sun falls below the minimum temperature required to sustain the thermonuclear burning process. Further, we demonstrated the fact that an event such as the thermonuclear burning of hydrogen into helium does not necessarily result in the loss of the mass of the sun. We speculate that the mass of the sun remains unaltered, even if it radiates billions of watts of energy per second. In Section 8.6, we explained the reasons for the opposite sense of the rotation for the planets Venus, Uranus and Pluto, on their axes, from the east to the west. The main reason is that their asymmetry of the mass distribution caused the gas molecules to the speed-up in a preferred direction of rotation.

In Section 8.7, we answered a puzzling question, why comets, such as Halley's Comet, has a very eccentric asymmetric orbit with a high orbital period of seventy-six years? It turns out that comets have a very long tail of gaseous matter. The excessive amount of gaseous matter shifts the center of the mass in a region outside of the solid matter. The unbalance of the center of mass, results in a sling effect when the comet experiences the gravity of the sun. This sling effect on the mass, throws it much further from the sun. In Section 8.8, we discussed the main purpose of the advanced string theory from Dr. Brian Greene. We indicated that additional dimensions are essential to explain the unknown propagation of the gravitational field's phenomena. However, it is not necessary that presence of an interacting particle, such as Higgs Boson, is crucial. One of the prime limitations of the string theory is that there is no evidence that the strings exists in nature. A major accomplishment of the string theory is that it blends the celestial mechanics of vast objects, stars and galaxies in our universe, with the smallest particles of quantum mechanics, quarks, through the vibrations of strings with great harmony.

The focus of the next chapter is how physicists bridge the gap between quantum mechanics and classical mechanics, and the purpose behind such unification. First, we shall revisit the results of Young's double slit experiment to explain the need for quantum mechanics in explaining the wave model of particles. Next, we shall discuss the blackbody radiation model proposed by Planck. After that, we shall look at the role of Dirac's equation in the formulation of quantum ideas. Then, we shall discuss the practical applications of quantum mechanics, such as electron tunneling microscopes, and computing the location of a quantum electron orbiting nucleus in the shells of atoms with well defined quantized states. We will end the chapter by discussing applications of quantum physics in future.

9.

Bridging Quantum Mechanics and Classical Mechanics

In this chapter, our focus is on how to connect the principles of quantum mechanics originally developed by Schrödinger in the 1920s and the principles of classical mechanics founded by Sir Isaac Newton in the 17th century. We shall demonstrate that one can unify the principles defined by Newton's Laws and Kepler's Law for predicting the behavior of large celestial objects, planets, stars, galaxies, and black holes, in the universe with the principles stated by Schrödinger, Planck, and Werner Heisenberg, for quantum (miniscule, nano) particles with the highest degree of success. You will soon realize that the incomplete ideas of relativity, developed by Einstein and others, may not be essential to bridge the gap between Newtonian mechanics and quantum mechanics.

In Section 9.1, we will revisit the behavior of a quantum particle and an interpretation of Quantum mechanics. Before we investigate details of double slit experiment, we shall define a distinct property of particle which distinguishes it from wave. For this purpose we state and prove particle wave separation theorem. Next, we shall examine the results of Young's double slit experiment and the Davisson-Germer experiment, named for Clinton Davisson and Lester Germer. In the double slit experiment, we will show that the conclusion, an electron interacts with both slits simultaneously in that experiment, is not vital for the success of the quantum particle model. Also, we will discuss some characteristics that distinguish the particle from the wave entities, such as the center of gravity, in more detail. In Section 9.2, we will describe the black body radiation. The discrete black body radiation model is often utilized to favor the particle and wave dual model for light waves. We are re-examining the black body radiation phenomena to prove that the correct discrete model proposed by Plank for the black body radiation does not imply that the energy radiation from the black body consists of particles. However, we emphasize that the model proposed by Plank is one of the greatest contributions that leads to the development of quantum ideas and mechanics.

In Section 9.3, we will focus on the contributions from yet another important physicist, Paul Dirac. It was the genius of Dirac, an expert mathematician and physicist from Great Britain, who essentially completed the effort, started by Max Plank and provided the equations describing the quantum particle behavior in its completeness. Sadly, many text books about physics have not acknowledged Dirac's contribution and have failed to recognize his work except that they have briefly mentioned his discovery of the positron, the anti-particle to the electron. Therefore, we have devoted this section to honor his work. We included the details of his equations and applications to explain other discoveries and theories, the anti-particle to the electron, the discovery of the positron and hole theory, Schrödinger's equations, Pauli's theory, Klein-Gordan's equations (named for Oskar Klein and Walter Gordon), and the energy per oscillator coupled to a heat bath, to some extent [33]. Obviously, to give total justification to his career may require a full text book which is outside the scope of this book. However, we shall show that Dirac's equation may be applied to investigate and analyze the superconductivity properties exhibited by certain materials. Since the superconducting phenomena plays a very important role in the development of the high speed transport system, this may lead to be an interesting analysis tool with which the worlds' physicists will play. Specifically, there has been a lot of research is in progress for developing the superconductive material at a higher temperature because at present, super conduction is possible only at a cryogenic temperature as low as 143°K.

In Section 9.4, we will describe the connection between Newton's classical mechanics and quantum mechanics, and show that the classical mechanics ideas may be extended to include quantum particles. In Section 9.5, we will discuss some of the existing applications of quantum mechanics, such as the scanning tunneling microscopes (STM), nano scale devices, and real-time computations for the position of the electrons in the atom shells, as well as new applications. One of the newer applications of quantum mechanics is the computations of the mean free path of the electrons and hydrogen ions in the thermonuclear burning of hydrogen into helium, a process occurring in the sun. As explained in Chapter 8, the computation of the mean free path plays an important role in the determination of the life span of the solar system which is vital for the survival of life on Earth. Finally, in Section 9.6, we will briefly discuss the future of quantum mechanics.

9.1 Revisiting Quantum Particle Behavior

The root of quantum mechanics lies in the experimental evidence that both matter and electromagnetic radiation are sometimes best modeled as particles and sometimes as waves, depending on the phenomenon observed. The evidence came from the experiments of the Davisson-Germer and Young's double slit [22]. We agree that the behavior of the wave or particle entities may be more suitably modeled as a particle or wave, depending on the situation observed. However, the model of behavior is different from prevailing reality. For instance, the diffraction patterns created by the electrons in the Davisson-Germer experiment are caused by the wave like motion of the electrons. This motion of the electron does not transform the electron particle into waves. Here, we are making a remarkable distinction between the particle nature and the wave nature of entities. For instance, the wave entities do not have a fixed location for the center of mass. Contrary to that, the particle entities can be characterized by a center of mass whose location varies with the position of the particle. The position of the center of mass stays fixed in relation to the mass distribution of an object or a particle of interest. Since the center of mass plays a very important role in deciding if an entity is a wave or a particle, we are formulating a theorem.

Theorem: Particle Wave Separation

Wave entities, the energy waves from the electromagnetic spectrum, does not have a rest mass for the transformation into energy and lacks the center of gravity in any moving or stationary frame of reference. These wave entities (light waves, sound waves, electromagnetic waves, X-rays, and UV-rays), without a center of mass, are not deflected and cannot be affected by the force of gravity.

This theorem clearly states the most distinguishing property between waves and particles. From this discussion, it will be obvious that the many scientific observations and interpretations of the results of experiments are highly subject to the performer's bias. We believe that the conclusion made in Young's double slit experiment, involving the passage of an electron through a double-slit, that an electron interacts with both slits simultaneously, was biased in favor of the wave nature [22]. What we are saying is that even if the act of measuring the motion of the electron, passing through a slit, destroys the interference pattern and the probability that an electron would pass through both slits is exactly zero. The probability will be non-zero only if the spacing between the two slits approaches zero, in which case it will be one slit. The valuable results for both experiments were that the dual complementary models, for both particles and waves, can successfully explain many of the phenomena that would not be understood by one type of model alone. Nevertheless, we agree that the wave model for particles and the de Broglie wavelength properties exhibited by particles are valid modeling concepts.

Now we shall discuss how one could apply the energy states of particles to unify the standard model for atoms and molecules whose behavior may be predicted using Newton's laws and the quantum particle model predicted by Schrödinger's laws of quantum mechanics. It is obvious that Newton's laws of motion flow smoothly for celestial objects of vastly large sizes, such as stars, galaxies, black holes, supernova, and nebulae. Also, the laws are obeyed by large objects on the Earth, such as buildings, ships, aircraft, and spacecrafts, down to objects the size of golf-balls. The laws hold well for matters in gas, liquid, and solid states. The laws are applied to much smaller sized objects comprised of molecules and atoms. It appears that Newton's laws breakdown at the subatomic particles, the protons, neutrons, and electrons, inside the nucleus of atoms. At this dimension, the weak Newtonian forces have hardly any effect on the motion of particles; as compared to the force because of the charges held within those particles (refer to Table 2.5). The failure of classical mechanics to model the behavior of the quantum particle stems from its inability to account for the local variations in the energy of the particle in a box. Because of the high speed of the electrons and the dimensions of the sub-nucleus, the distances are so small that finding the exact location of the electron particle in the nucleus is a difficult task by applying the standard principles of classical mechanics. The Heisenberg uncertainty principle applies unfavorably to the problem and makes the situation worse. Therefore, it makes sense to develop a model for the quantum particle electron that can precisely find the position of the electron in atomic orbits.

In 1928, Max Born extended the idea of the de Broglie wave model associated with a particle to find the probability density function of a particle. The relative probability, per unit of volume that the particle will be found at any given point in the volume is:

$$\mathbf{P(x, y, z)\ dV = |\Psi^2|\ dV} \qquad \mathbf{(9.1)}$$

where Ψ is called probability amplitude or wavefunction

This equation was derived by the premise that the probability of finding a photon (assume a photon is a particle for this purpose, the particle may be neutrino) in space [22] is:

$$\mathbf{P_{photon}(x, y, z)\ dV = E^2\ dV} \qquad \mathbf{(9.2)}$$

where **E** is the electric field amplitude

In 1926, Schrödinger proposed a wave equation that describes the manner in which the wavefunction Ψ changes in space and time. One of the major applications of the Schrödinger equation is to calculate the quantum amplitudes of the probability of distribution as a function of the position and time for the particles of a given type in a physical environment, for instance, electrons at the interior of an atom. Since our intention is to analyze the results, we shall skip the details of derivation [3] and state the equations:

$$-\hbar^2/(2\ m)\ \partial^2\Psi/\partial x^2 + V\Psi = E\Psi \qquad (9.3)$$

and

$$-\hbar^2/(2\ m)\ \partial^2\Psi/\partial x^2 + V\Psi = i\hbar\ \partial\Psi/\partial t \qquad (9.4)$$

These equations are two different forms of Schrödinger's equations in one dimension. The former equation, that does not contain the time explicitly, is known as the time-independent form and is most frequently used to analyze the stationary states of atomic systems. The stationary states of Bohr's theory are quantized for a bound particle with discrete energy values. The solution for the one dimensional, time-independent form of Schrodinger's equation provides for the analysis of the characteristic frequencies for the normal mode of the system. The time-dependent form must be used when dealing with problems involving the actual motion of particles from one point to another. For instance, the time-dependent solution form may be used to locate the position of an electron inside the atom structure of any element on the periodic table at any particular time. It is discovered that by applying Schrodinger's equations and the principles of quantum mechanics tunneling of the particle through a potential energy barrier may be analyzed successfully [22]. The phenomena could not be explained by applying the treatment of classical mechanics because it forbids the penetration of a particle inside a potential energy barrier whose height is above the particle's energy level. An excellent treatment of Schrödinger's equation for a particle in a three dimensional box and the solutions for complex problems, such as spherically symmetric potential in hydrogen nucleus, may be found from [3] *An Introduction to Quantum Physics*, Anthony French and Edwin Taylor. By applying Schrödinger's equations to the quantum particle electron, one can precisely compute the probability of finding the particle at any point and at any time, even in the presence of gravitational effects. Therefore, Schrödinger's equations provide a means to connect the mechanics of large celestial objects with microscopic nucleus particles, without any discrepancy.

In the next section, we will discuss another important topic that formed the basis for the foundation of the quantum theory by German physicist, Max Planck, who quantized the model for heat and radiation energy spectral distribution from a black body at a high temperature. He was awarded the Nobel Prize in physics for his discovery of the quantum nature of energy. We are in complete agreement with Planck's hypothesis and the discrete energy model for black body radiation. However, we disagree with Einstein for extrapolating the principles laid out by Planck and applying it to the situation for the cavity oscillations of the electromagnetic field. He stated that the oscillations were quantized themselves, to the extent that quantization leads to the photon particle nature of light. The topic of the black body radiation played a very important role in establishing the fact that light and other radiation energy waves, such as infrared, X-rays, and ultraviolet rays, possess the quantized energy levels and may be characterized as such. Further, this important discovery lead Einstein to believe that indeed, energy waves such as light waves, may consist of particles of photons. Subsequently, he invented the photoelectric effect to convince physicists that light may have a dual nature. During our discussion with Professor Steven Manly, from the University of Rochester, Rochester, New York, he raised a good question. He asked if we could explain the black body radiation by the classical model. In that case, he told the world physics community that he will gladly accept the wave-only nature of light. After reading this chapter, you will be convinced that we do not contest the quantized model for the energy levels in quantum oscillations inside a black body or, for that matter, any kind of radiation. However, we do state that the quantized level for energy content in the waves need not imply that the waves consist of photon particles. Since an extensive study of the quantum theory is beyond the scope of this book, we shall touch on the very basic principles of quantum ideas born from Planck's hypothesis.

9.2 Black Body Radiation and Quantum Ideas

Let us first explain why physicists in the 1900s believed that a different theory or model, other than classical physics, is required. One of the reasons was that they were not able to explain and model the behavior of matter at the atomic scale. For instance, the emission of discrete wavelengths of light and the radiation from atoms of a gas at high temperature could not be explained by the principles of classical physics. Consider an example of the heat radiation and absorption from a black body. A close approximation for the black body is a hollow object with a small hole. Any radiation incident, in the hole from outside, enters the cavity. The radiation is absorbed by the interior walls and reflected multiple times, hence, the object with a cavity looks like a black body, as indicated in Figure 9.1. Careful studies show that the radiation escaped from the hole of the cavity consists of a continuous distribution of wavelengths encompassing the whole section of the electromagnetic spectrum. A characteristic of this radiation is that it depends on the absolute temperature of the body and does not depend on the kind of material used to make the body.

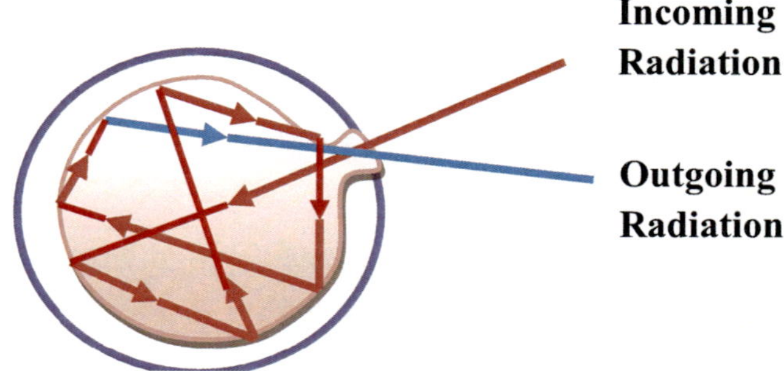

Figure 9.1 Simulated black body radiation. An opening to a cavity inside a hollow object is a good approximation for black body. The energy radiated from the opening is characterized by the temperature of the object, not on the material. The escaped radiations reflect the standing wave modes of radiation created by multiple reflections. (Courtesy Serway & Jewett [22]).

Further studies show that the radiation emitted by oscillations in the cavity walls, experience multiple reflections inside the cavity. Standing waves of radiation energy are established within the cavity. The distribution of the energy among these standing wave modes determines the distribution of the radiation leaving the cavity through the hole. Figure 9.2 illustrates how the intensity of the black body radiation varies with the temperature and wavelength. The intensity vs. wavelength curves is plotted for three different values of temperature, **2000°K**, **3000°K**, and **4000°K** respectively.

The following two findings were of considerable importance:

1. The total power of the emitted radiation increases with the fourth power of temperature, known as Stefan-Boltzmann's law (named for Jožef Stefan and Ludwig Boltzmann) of black body radiation in thermodynamics.

   $$\mathbf{P = \sigma A e T^4} \qquad \mathbf{(9.5)}$$

   where **P** is the power in watts radiated at all wavelengths from surface of an object
   σ is the Stefan-Boltzmann constant equal to $\mathbf{5.67 \times 10^{-8}\ W/m^2\ K^4}$
   A is the surface area of the object in square meters
   e is the emissivity of the surface
   T is the surface temperature in °**K**

   for black body **e = 1** precisely.

2. The peak of the wavelength distribution shifts to shorter wavelengths as the temperature increases. The relationship to describe this finding is called Wien's displacement law:

$$\lambda_{max}T = 2.898 \times 10^{-3} \text{ m K} \quad (9.6)$$

where λ_{max} is the wavelength at which the curve peaks
T is the absolute temperature of the surface emitting radiation

Wien's displacement law, named for Wilhelm Wien, is consistent with the observable behavior of the heated black body. At room temperature and low temperatures, the object does not glow though it emits radiation in the infrared region of the spectrum due to the temperature differential between it and the ambient environment. At a slightly higher temperature, 3000 K, it glows red because the peak shifts to the near-infrared visible end of the spectrum red color. At a still higher temperature, approximately 6000 K, it glows white because the peak is shifted in the all visible light color spectrum so that all colors are emitted producing white light.

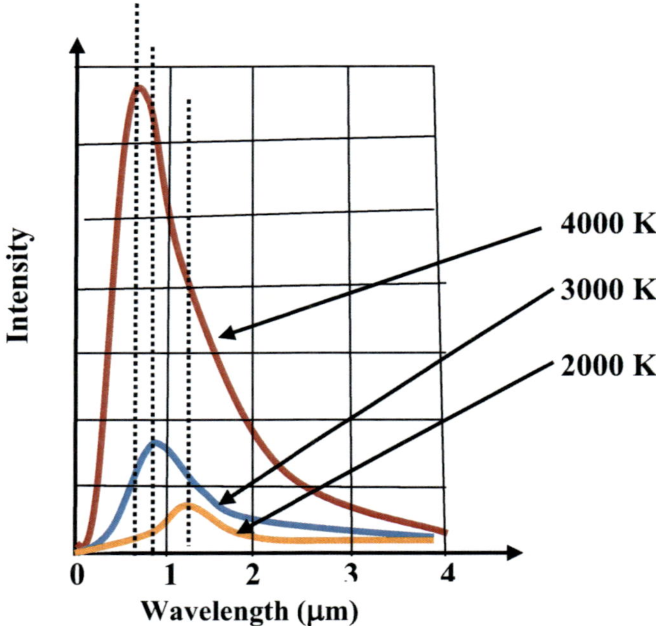

Figure 9.2 Intensity of black body radiation versus wavelength at three temperatures. The visible wavelength is **0.4 μm** to **0.7 μm**. Notice how the peak intensity wavelength shifts toward smaller wavelengths for high temperatures. At **3000°K**, the peak wavelength is slightly above the infrared (**07 μm**) region such that the object glows red. At approximately **6000°K**, the peak is shifted to **0.4 μm** wavelength at which all color is radiated forming white light. (Courtesy Serway & Jewett [22])

A successful theory for black body radiation must at least satisfy the requirements of equations (9.5) and (9.6). It turns out that the classical model to describe the distribution of energy from the black body, Rayleigh-Jeans law (named for John William Strutt, 3rd Baron Rayleigh, and Sir James Jeans) fails to meet the two requirements at low frequencies. The expression for the intensity or power per unit area emitted is based on classical theory, Rayleigh-Jeans law, as follows:

$$\mathbf{I(\lambda,T) = 2\pi c k_B T/\lambda^4} \qquad \textbf{(9.7)}$$

where $\mathbf{k_B}$ is the Stefan-Boltzmann constant
which is equal to $\mathbf{5.67 \times 10^{-8}\ W/m^2\ K^4}$

The value of the intensity from this function approaches infinity as the wavelength λ approaches zero. This is a major disagreement with the observed data illustrated in Figure 9.3. According to the measured data, as the wavelength approaches zero, the intensity for the wavelength also reduces to zero. The mismatch between theory and the experiment was so disconnected that the scientists called it an ultraviolet catastrophe.

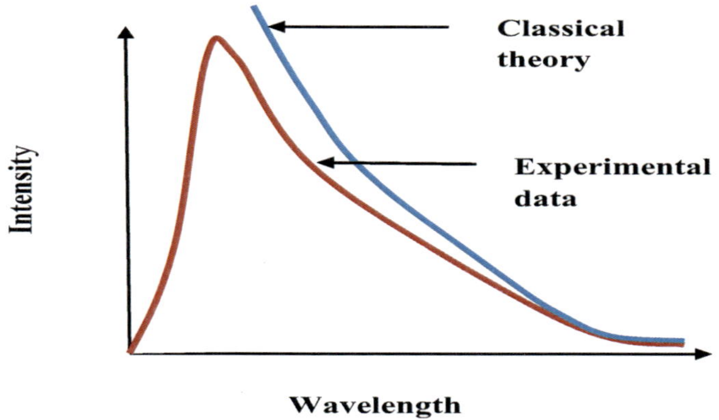

Figure 9.3 Experimental results of black body radiation. The curve predicted by the Rayleigh-Jeans law of classical theory shows discrepancy at small wavelengths. (Courtesy Serway & Jewett [22])

After the catastrophe in 1900, Planck developed a theory of the black body radiation that was in total agreement with the experimental results at all wavelengths. He made two bold assumptions about the nature of the oscillations in the cavity walls:

- The energy of a particle in the oscillator can have only certain discrete values:

$$\mathbf{E_n = n \times h \times f} \qquad \textbf{(9.8)}$$

where **n** is a positive integer known as quantum number
f is the frequency of oscillations
h a constant introduced by Max Planck, known as Planck's constant

- The oscillator or the particle involved in the oscillator emits or absorbs the energy when making a transition from one quantum state to the other. The entire energy difference between the starting and final states in the transition is emitted or absorbed as a single quantum of radiation. A particle in the oscillator will emit or absorb energy only when it changes the state.

$$\mathbf{E = h \times f} \qquad \textbf{(9.9)}$$

He turned out to be correct in those assumptions. His two predictions were the key points and highlights of Plank's career and the theory of black body radiation. He was awarded the Nobel Prize for this remarkable discovery. During his time, many scientists did not believe that the quantum ideas from Plank were realistic. They, including Plank himself, thought that the computations were correct by a pure coincidence. As we shall soon discover, the major difference between Plank's model and the Rayleigh-Jeans classical model was in the method of computation of the average energy associated with a particular wavelength of the standing waves in the cavity. For a detailed explanation of the theory of black body radiation, see [22] Serway, R. and Jewett, J., *Physics for Scientists and Engineers with Modern Physics*. Figure 9.4 illustrates the allowed energy levels for an oscillator inside the cavity of a black body with frequency **f.** We shall state Plank's theoretical expression, which he derived with the same classical ideas proposed by the Rayleigh-Jeans model, after correction for intensity computations at small wavelengths:

$$\mathbf{I(\lambda,T) = 2\pi\, hc^2/\lambda^5(e^{hc/kBT} - 1)} \qquad \textbf{(9.10)}$$

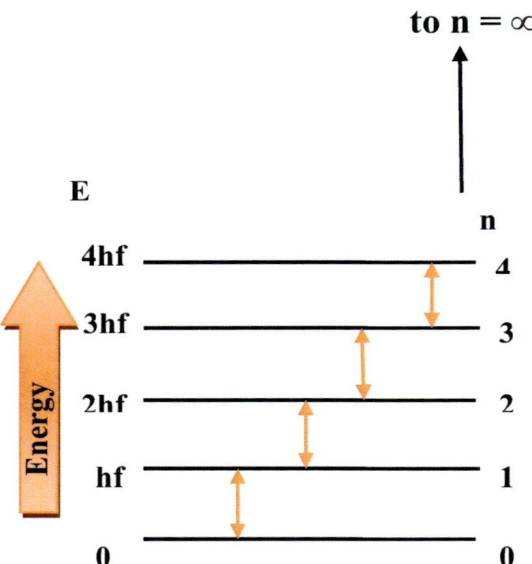

Figure 9.4 Allowed energy levels for an oscillator with frequency f. The allowed transitions are quantized (discrete values in steps of hf) as indicated by double-headed arrows.

In the Rayleigh-Jeans model, the average energy associated with a specified wavelength of the standing waves in the cavity, is the same for all wavelengths and is equal to $\mathbf{k_B T}$. In Plank's model, a wave's average energy is the average energy difference between the levels of the oscillator, weighted according to the probability of the wave being emitted. The weighting is based on the occupation of higher energy states, as described by the Boltzmann distribution law. The probability of a state being occupied is proportional to the factor $\mathbf{e^{-E/kBT}}$. In equation (9.10), Plank adjusted the value of the parameter **h** such that the curve from the expression matched the data at all wavelengths.

The value of his constant **h** is:

$$\mathbf{h = 6.626 \times 10^{-34}\ J.S}$$

Figure 9.5 illustrates that the curve presented by Plank's model fits the data for the entire range of wavelengths with total consistency.

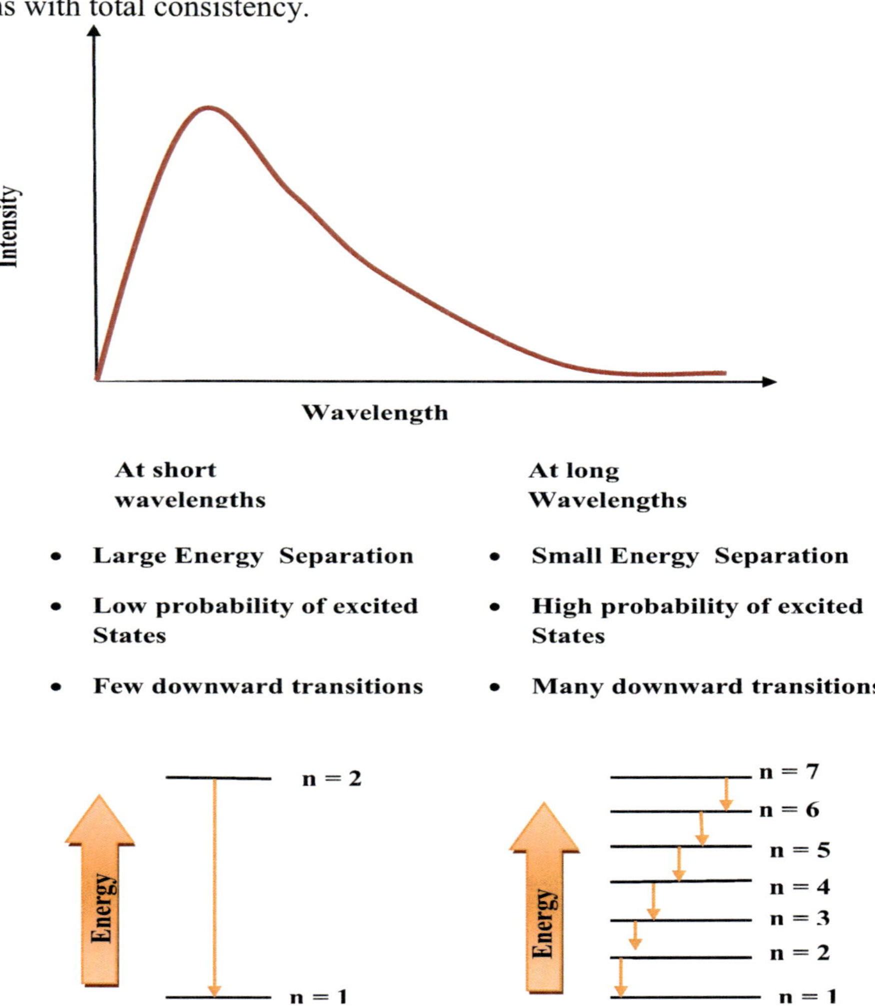

Figure 9.5 Plank's model: Intensity vs. Wavelength. In his model, the average energy associated with a given wavelength is the product of the energy of a transition state and a factor related to the probability of transition occurring. (Courtesy of Serway & Jewett [22].)

After looking at the success of Plank's theory, in 1905 Einstein re-derived Plank's results assuming that the cavity oscillations of the electromagnetic field themselves were quantized. He proposed that the quantization is a fundamental property of light and other electromagnetic radiation. To us this is not a surprise because the radiations are a by-product of the change in the quantum state of the quantum particle electron in the spherical atom shell of material of the cavity, as per Planck's hypothesis. Therefore, light and other radiation inherits the quantized energy state property from its parent particles, as stipulated by Plank's black body radiation model and is complemented by Pauli's exclusion principle of the standard model for atoms. The credit for the marvelous exclusion principle goes to a great Austrian theoretical physicist, Wolfgang Pauli. His exclusion principle states that:

No two electrons can ever be in the same quantum state; therefore, no two electrons in the same atom can have the same set of quantum numbers.

When Pauli stated the same quantum state, he implied that the energy level possessed by the electron in that quantum state is unique for an atom of the element from which the matter was formed. This unique quantization of the energy state, results in the unique energy level for the light or radiation energy emission when an electron orbit transition occurs. Therefore, even though the energy content of the light waves is quantized, the conclusion should not be applied in the pursuit of the light wave as the photon particle model. Our reasoning to oppose characterizing the light wave as the photon particle stems from the fact that the light energy waves do not have a static location for the center of gravity. The unique property of the particle is that it should have a unique position for its center of gravity at all times, this completely distinguishes it from wave entities. Therefore, you will be convinced and support our views that the light and energy waves are not comprised of any particles.

In the next section, we will discuss a very valuable contribution to quantum mechanics from another great physicist, Dirac [32], a student at Bristol University in London between 1918 and 1923. First, he was a student in electrical engineering and then in applied mathematics. Dirac became famous when he derived the wave equations for the quantum particle electron that had four components. We shall understand the intriguing details of his equations next. At this time, it is almost certain that any quantum physicist cannot afford to escape learning the tricks presented by Dirac. His equations lead to the discovery of a series of anti-matter particles, positrons, the antiparticles to electrons, anti-protons, and anti-quarks. Eventually, he was recognized as the father of modern quantum electrodynamics (QED), when in September 1997, Dirac House, the headquarters of the Institute of Physics Publishing, was officially opened in London, UK.

9.3 Role of Dirac in Quantum Electrodynamics

Before we describe the exact details of Dirac's equations, let us look at the history and background events, in brief, which prompted Dirac to come up with his fascinating ideas and equations. By his nature, Dirac was a very quiet man and was not known outside of physics because his work was highly mathematics-oriented. That is one of the reasons why seldom can anyone find diagrams or figures in any of the text books that describe his work. When Dirac was a student at Cambridge University in 1923, the physics of atoms was in its infancy. At that time, many physicists were astonished to discover that the light coming from atoms had energy that is quantized, meaning it had values that are described in constant increments. Classical mechanics would predict that the energy value should be any number in a continuous set of numbers. Soon after that discovery, Dirac realized a need for developing new equations and principles of science which explain the behavior of the particles inside the nucleus of atoms. He looked for solutions in physics, of particles that are small, yet moving at a high the speed, nearing the speed of light. He noticed that if one constructs the quantum theory that is consistent with relativity, it would result in probabilities with negative numbers, which they were not able to explain then.

To resolve this mentioned dilemma, he asked questions, such as what is the right kind of quantum equations or models that would describe the electron energy states in orbit? What are the wave equations that govern the dynamics of those waves which also satisfy the requirements of relativity? After the invention of the uncertainty principle by Werner Heisenberg in 1925, the quantum rules emerged automatically by a peculiar framework. In Dirac's view, the object that represents the variables one can measure in the experiments should be treated as operations. In such a system, the outcome of an operation, such as multiplication, would depend on the order in which that operation is performed on the variables of interest. The same unification was found to agree with Schrödinger's quantum mechanics, where the state of the system is represented by a wave whose strength gives the probabilities of the different results of measurements. In that framework, Dirac was able to show that the simplest wave equation describing the motion of an electron had four components. The computed energy radiated values from the electrons in orbit agreed with the precise measurements of light emitted from atoms.

Any great physicist's theory is conceived in order to provide more sense when it solves the problem that inspired its construction, and it explains more facts and predicts more new things. This happened naturally in Dirac's case. In Dirac's equation, a new fact that emerged was the electron spin that need not be imported. The magnetism associated with the electron that was both a quantum particle and a relativistic one, revealed the spin moment. Another interesting aspect of his equation was that the solution provided by the equation would always have a solution in a pair, a positive energy part and a counter-part negative energy. He interpreted the counter-part solution to claim existence of the counter-particle. For instance, the positron was recognized as the counter-particle to the electron that had the same mass as an electron and a charge of the same magnitude with opposite polarity. Dirac's work was applied to explain how light interacts with matter, and in-turn, how atoms radiate the light waves with precise energy levels. Beyond that, Dirac predicted that when the particle electron encounters the anti-particle positron, the charges on both particles neutralize and their mass is transformed into radiation energy. The process was known as annihilation and thus, anti-matter's existence was predicted. When Anderson discovered and verified the existence of the positron, he was awarded the Nobel Prize in 1932. Now, we will describe the exact details of Dirac's equation.

In physics, the Dirac equation is a relativistic quantum mechanical wave equation, formulated in 1928, that provides a description of the elementary spin particles, such as electrons, consistent with both the principles of quantum mechanics and the special theory of relativity. The equation demands the existence of antiparticles and actually predated their experimental discovery, making the discovery of the positron, the antiparticle of the electron, one of the greatest triumphs of modern theoretical physics.

Dirac started with an equation that was similar to the Schrödinger for a free particle:

$$-\frac{\hbar^2}{2m}\nabla^2\phi = i\hbar\frac{\partial}{\partial t}\phi. \tag{9.11}$$

The left side represents the square of the momentum operator, divided by twice the mass, which is the non-relativistic kinetic energy. If one wants to get a relativistic generalization of this equation, then the space and time derivatives must be entered symmetrically, as they do in the relativistic Maxwell; the derivatives must be of the same order in space and time. In relativity, the momentum and the energy are the space and time parts of a geometrical space-time vector, the 4-momentum, and they are related by the relativistic invariant relation:

$$\frac{E^2}{c^2} - p^2 = m^2c^2 \tag{9.12}$$

Which says that the length of this vector is the rest mass ***m***. Replacing ***E*** and ***p*** by $i\hbar\frac{\partial}{\partial t}$ and $-i\hbar\nabla$, as Schrödinger's requires, we get a relativistic equation:

$$\left(\nabla^2 - \frac{1}{c^2}\frac{\partial^2}{\partial t^2}\right)\phi = \frac{m^2c^2}{\hbar^2}\phi \tag{9.13}$$

and the wavefunction ϕ is a relativistic scalar: a complex number which has the same numeric value in all frames. Because the equation is the second order in the time derivative, one must specify both the initial value of $\partial_t\phi$ and not just ϕ. This is normal for classical waves where the initial conditions are the position and the velocity. However, in quantum mechanics, the wavefunction is supposed to be the complete description; just knowing the wavefunction should determine the future. Dirac was thinking in terms of the wavefunctions rather than the fields, he reasoned that what was needed was an equation that was first-order in both space and time. One could formally take the relativistic expression for the energy:

$$E = c\sqrt{p^2 + m^2c^2}$$

and replace p by its operator equivalent, expand the square root in an infinite series of derivative operators, set up an eigenvalue problem, and then solve the equation formally by iterations. Most physicists had little faith in such a process, even if it were technically possible. But for Dirac this was not the case. Therefore, when Dirac was staring into the fireplace at Cambridge University pondering this problem, he hit upon the idea of taking the square root of the wave operator on the left side of Schrödinger's equation, thus:

$$\nabla^2 - \frac{1}{c^2}\frac{\partial^2}{\partial t^2} = (A\partial_x + B\partial_y + C\partial_z + \frac{i}{c}D\partial_t)(A\partial_x + B\partial_y + C\partial_z + \frac{i}{c}D\partial_t) \quad \textbf{(9.14)}$$

On multiplying out the right side, we see that in order to get all of the cross-terms, such as $\partial_x\partial_y$, to be consumed, we must assume:

$$AB + BA = 0, \ \ldots \quad \textbf{(9.15)}$$

and

$$A^2 = B^2 = \ldots = 1 \quad \textbf{(9.16)}$$

Dirac, who had just then been intensely involved with working out the foundations of Heisenberg's matrix mechanics, immediately understood that these conditions could be met if **A**, **B** are *matrices*, with the implication that the wavefunction has *multiple components*. This immediately explained the appearance of the two-component wavefunctions in Pauli's phenomenological theory of spin, something that up until then had been regarded as mysterious, even to Pauli himself. However, one needs at least **4×4** matrices to set up a system with the properties desired, so, the wavefunction had *four* components, not two, as in the Pauli theory.

Given the factorization in terms of these matrices, one can immediately write an equation:

$$(A\partial_x + B\partial_y + C\partial_z + \frac{i}{c}D\partial_t)\psi = \kappa\psi \quad \textbf{(9.17)}$$

with κ to be determined

Applying again the matrix operator on either side yields:

$$(\nabla^2 - \frac{1}{c^2}\partial_t^2)\psi = \kappa^2\psi \tag{9.18}$$

On taking $\kappa = mc/\hbar$, we find that all the components of the wavefunction individually satisfy the relativistic energy–momentum relation. Thus, the sought-for equation that is first-order in both space and time is:

$$(A\partial_x + B\partial_y + C\partial_z + \frac{i}{c}D\partial_t - \frac{mc}{\hbar})\psi = 0 \tag{9.19}$$

with $(A, B, C) = i\beta\alpha_k$ and $D = \beta$, we get the Dirac equation

The Dirac equation, formally proposed by Dirac, is as follows:

$$\left(\beta mc^2 + \sum_{k=1}^{3} \alpha_k p_k \, c\right) \psi(\mathbf{x}, t) = i\hbar \frac{\partial\psi(\mathbf{x}, t)}{\partial t} \tag{9.20}$$

where **m** is the rest mass of the electron
c is the speed of light
p is the momentum operator
x and *t* are the space and time coordinates

A more general form of equation may include **y** and **z** components of function **ψ:**

$\hbar$ is the reduced Planck's constant, also known as Dirac's constant

$$\hbar = \frac{h}{2\pi} \tag{9.21}$$

The new elements in this equation are the **4×4** matrices, α_k and β, and the four-component wavefunction ψ. The matrices are all Hermitian and have squares equal to the identity matrix:

$$\alpha_i^2 = I_4 \tag{9.22}$$

$$\beta^2 = I_4 \tag{9.23}$$

They all mutually anti-commute:

$\{\alpha_i, \alpha_j\} = 0$ **and** $\{\alpha_i, \beta\} = 0$.

Explicitly:

$$\alpha_i \alpha_j = -\alpha_j \alpha_i, \quad \textbf{(9.24)}$$

$$\alpha_i \beta = -\beta \alpha_i, \quad \textbf{(9.25)}$$

where ***i*** and ***j*** are distinct and range from **1** to **3**

These matrices, and the form of the wavefunction, have a deep mathematical significance. The algebraic structure represented by the Dirac matrices, had been created some 50 years earlier by the English mathematician William Kingdon Clifford, which in turn had been based on the mid-19th century work of the German mathematician Hermann Grassmann in his "Lineare Ausdehnungslehre" (Theory of Linear Extensions). The latter had been regarded as well-nigh incomprehensible by most of his contemporaries. The versatility of Dirac's equation comes from the fact that it could explain many theories by applying the appropriate mathematics treatment to his fundamental equations. For instance, one could shed light on Pauli's theory and the hole theory by deriving the field equations from Dirac's equation. Also, it can explain many effects after comparing his equations with the corresponding equations for the effect. The examples are the Schrödinger equation, the Klein-Gordon equation, coupling to an electromagnetic field, the Lorentz relativistic invariance, and the solution of field equation in curved space-time. The complete analysis of these issues is beyond the scope of this book [33]. We shall illustrate this fact by citing the details of identification of the observable phenomena (9.26).

The Dirac theory, while providing a wealth of information that is accurately confirmed by experiments, nevertheless introduces a new physical paradigm that appears, at first, difficult to interpret and even paradoxical. Some of these issues of interpretation must be regarded as open questions. The critical physical question in a quantum theory is, what are the physically observable quantities defined by the theory? According to general principles, such quantities are defined by the Hermitian operators that act on the Hilbert space of the possible states of a system. The eigenvalues of these operators are then the possible results of measuring the corresponding physical quantity. In the Schrödinger theory, the simplest such object is the overall Hamiltonian, which represents the total energy of the system. If we wish to maintain this interpretation on passing to the Dirac theory, we must take the Hamiltonian to be:

$$H = \gamma^0 \left(mc^2 + c \sum_{k=1}^{3} \gamma^k (p_k - \frac{q}{c} A_k)\, c \right) + qA^0 \quad \textbf{(9.26)}$$

This looks promising, because we see by inspection, the rest energy of the particle and, in case $\boldsymbol{A = 0}$, the energy of a charge placed in an electric potential $\boldsymbol{qA^0}$. What about the term involving the vector potential? In classical electrodynamics, the energy of a charge moving in an applied potential is:

$$H = c\sqrt{(p - \frac{q}{c}A)^2 + m^2c^2} + qA^0 \quad \textbf{(9.27)}$$

Thus, the Dirac Hamiltonian is fundamentally distinguished from its classical counterpart and we must take great care to correctly identify what is an observable in this theory. It is evident that much of the apparent paradoxical behavior implied by the Dirac equation, amounts to a misidentification of these observables. The complete analysis of the phenomena, using Dirac's equation, requires an advanced knowledge of mathematics that we have refrained from applying.

Since the Dirac equation was originally invented to describe the electron, we will generally speak of "electrons" in this book. The equation also applies to quarks, which are elementary spin **-½** particles, too. A modified Dirac equation can be used to approximately describe the protons and neutrons, which are not elementary particles, they are made up of quarks, but have a net spin of ½. Another modification of the Dirac equation, called the Majorana equation, named for Ettore Majorana, is thought to describe neutrinos, also spin **-½** particles.

The Dirac equation describes the probability amplitudes for a single electron. This is a single-particle theory; in other words, it does not account for the creation and destruction of the particles, and for the ultimate need to switch from the Dirac equation for wavefunctions to the physically distinct Dirac equation for fields. It gives a good prediction of the magnetic moment of the electron and explains much of the fine structure observed in the atomic spectral lines. It also explains the spin of the electron. Two of the four solutions of the equation correspond to the two spin states of the electron. The other two solutions make the peculiar prediction that there exists an infinite set of quantum states in which the electron possesses negative energy. This strange result led Dirac to predict, via a remarkable hypothesis known as the "hole theory," the existence of particles behaving like positively-charged electrons. Dirac at first thought these particles might be protons. He was chagrined when the strict prediction of his equation, which actually specifies the particles of the same mass as the electron, was verified by the discovery of the positron in 1932. When asked later why he had not boldly predicted the yet unfound positron with its correct mass, Dirac answered, "pure cowardice!" He shared the Nobel Prize in 1933 anyway.

Despite these successes, Dirac's theory is flawed by its neglect of the possibility of creating and destroying the particles, one of the basic consequences of relativity. This difficulty, and others, is resolved by reformulating it as a quantum field theory which involves giving up the notion of the wavefunction. Adding a quantized electromagnetic field to this field theory, leads to the theory of quantum electrodynamics (QED). Moreover, the equation cannot fully account for the particles of negative energy, but is restricted to the positive energy particles. A similar equation for spin **3/2** particles is called the Rarita-Schwinger equation, named for William Rarita and Julian Schwinger. In the next section, we shall describe a method to extend the principles of classical mechanics to connect it with quantum mechanics, so that the union between the principles of classical mechanics and the principles of quantum mechanics is possible. This marriage is important because it allows an explanation for all of the phenomena observed in the universe without losing the connection between the theoretical model of the most fundamental particles' behavior and the reality. For instance, when one models a particle as a wave, the connection between the model of the particle and the real particle is lost because the wave model lacks the information about the center of mass for the particle.

9.4 Connecting Classical Mechanics and Quantum Mechanics

In general, the principles of classical mechanics imply that the future of any particle system is predictable. One can precisely determine the trajectory of an object with the information for its starting position, velocity, and the forces exerted on it, with arbitrarily high precision. Let us consider an example of an event in which the alpha particles escape from the He nucleus via emission. This phenomenon, such as the radioactive decay of the helium nucleus, caused by the alpha particle emission, cannot be explained by classical mechanics. The reason is that the alpha particle cannot classically escape from the unstable, heavy He nuclei because it has to overcome the potential barrier from the nucleus maintained by the combination of attractive and repulsive charge forces which are several times larger than the energy of alpha particle. This escape of the alpha particle can be explained more readily by quantum mechanics. The primary strength of the quantum particle model comes from its ability to predict the future positions of particles with a definite probability. The environment of quantum mechanics adds flexibility in computing the future events based on the past information of the particle described by its wavefunction model. The probability-based model of the alpha particle systems is more efficient than classical mechanics that forbids the tunneling of the alpha particle and the escape for emission from the He nucleus.

Because we are proponents of the wave model for radiation energy, such as the alpha particle (alpha waves), the probabilistic quantum wave model is much more suitable to describe the emission event release of the energy waves and the decay of the He nucleus, than stating the escape of the alpha particles. Thus, quantum mechanics, in describing the alpha wave emission from the He nucleus, is much more realistic. The question is why the alpha waves are released from the He nucleus even though the potential barrier of the charges to the energy waves prevails inside the nucleus? The answer comes from the fact that the charge forces cannot trap the energy waves inside the nucleus. Whenever the energy waves are created inside the nucleus of any atom, from the energy state transition event of the fundamental particles electrons, the waves will escape regardless of the force from the charges. This is the reality that is consistent with stating that the alpha particles have tunneled through the potential barrier.

Let us look at another example of the fusion of protons with neutrons, a process of the thermonuclear burning of hydrogen into helium atoms occurring inside the sun. We saw a lot of details about the process in Section 8.5. Here, we will explain how two protons (hydrogen ions) are fused together in spite of the strong positive charges on both protons that repel into a proton and neutron stable structure of the helium isotope. Under normal or elevated temperature, the protons of the hydrogen ions do not fuse together as they can't overcome the repulsive force of the positive charge. However, under extremely high temperature and pressure, it may be possible that the force effect from the quarks within the two protons may dominate the overall charge force of the entire proton because the quarks of the two protons come too close. Once that event occurs, the quarks bind two protons together into a stable structure of a proton and a neutron. This fusion process can be more easily understood if the behavior of the protons and neutrons, and the associated quarks, is modeled using the quantum particle model. Since quarks do not exist freely in nature, it makes sense to apply the wave model to predict the behavior of quarks and quark-like particles, such as leptons, baryons, fermions, and bosons, as there is no loss of reality. Let us examine the details of one more example to indicate that Dirac's equation does provide a means to connect classical mechanics with quantum mechanics. We will achieve our objective by analyzing the details of Pauli's theory.

The roots of Dirac's equation stem from the necessity of introducing the half-integral spin momentum for the quantum particle free electron. The half-integral spin momentum property dates back, experimentally, to the results of the Stern–Gerlach, named for Otto Stern and Walther Gerlach. In the experiment, a beam of atoms is run through a strong inhomogeneous magnetic field, which then splits into N parts, depending on the intrinsic angular momentum of the atoms. It was found that for silver atoms, the beam was split in two –the ground states, therefore, could not be integral, because even if the intrinsic angular momentum of the atoms were as small as possible, 1, the beam would be split into three parts, corresponding to atoms with $\mathbf{L_z} = -\mathbf{1}, \mathbf{0}$, and $+\mathbf{1}$. The conclusion is that the silver atoms have net intrinsic angular momentum of ½. Pauli, with his genius, developed a theory which explained this splitting by introducing a two-component wavefunction and a corresponding correction term in the Hamiltonian, representing a semi-classical coupling of this wavefunction to an applied magnetic field, as such:

$$H = \frac{1}{2m}(\sigma \cdot (p - \frac{e}{c}A))^2 + eA^0 \tag{9.28}$$

Here $\boldsymbol{A}^{\mu}$ is the applied electromagnetic field, and the three sigma's are Pauli matrices. $\boldsymbol{e}$ is the charge of the particle, for example, $\boldsymbol{e} = -\boldsymbol{e}_0$ for the electron. On squaring the first term, a residual interaction with the magnetic field is found, along with the usual Hamiltonian of a charged particle interacting with an applied field:

$$H = \frac{1}{2m}(p - \frac{e}{c}A)^2 + eA^0 - \frac{e\hbar}{2mc}\sigma \cdot B \tag{9.29}$$

This Hamiltonian is now a $\mathbf{2} \times \mathbf{2}$ matrix, so the Schrödinger based on it:

$$H\phi = i\hbar\frac{\partial\phi}{\partial t} \tag{9.30}$$

must use a two-component wavefunction, Pauli had introduced the sigma matrices:

$$\sigma_k = \begin{pmatrix} 0 & 1 \\ 1 & 0 \end{pmatrix}, \begin{pmatrix} 0 & -i \\ i & 0 \end{pmatrix}, \begin{pmatrix} 1 & 0 \\ 0 & -1 \end{pmatrix} \quad \textbf{(9.31)}$$

as pure phenomenology. Dirac Now had a theoretical argument that implied that the spin was somehow the consequence of the marriage of the quantum theory to relativity. The Pauli matrices share the same properties as the Dirac matrices – they are all Hermitian, square to **1**, and anti-commute. This allows one to immediately find a representation of the Dirac matrices in terms of the Pauli matrices:

$$\alpha_k = \begin{pmatrix} 0 & \sigma_k \\ \sigma_k & 0 \end{pmatrix} \quad \textbf{(9.32)}$$

$$\beta = \begin{pmatrix} I_2 & 0 \\ 0 & -I_2 \end{pmatrix} \quad \textbf{(9.33)}$$

The Dirac equation may now be written as an equation coupling the two-component spinors:

$$\begin{pmatrix} mc^2 & c\sigma \cdot p \\ c\sigma \cdot p & -mc^2 \end{pmatrix} \begin{pmatrix} \phi_+ \\ \phi_- \end{pmatrix} = i\hbar \frac{\partial}{\partial t} \begin{pmatrix} \phi_+ \\ \phi_- \end{pmatrix} \quad \textbf{(9.34)}$$

Notice that on the diagonal, we find the rest energy of the particle. If we set the momentum to zero, that is, bring the particle to rest, and then we have:

$$i\hbar \frac{\partial}{\partial t} \begin{pmatrix} \phi_+ \\ \phi_- \end{pmatrix} = \begin{pmatrix} mc^2 & 0 \\ 0 & -mc^2 \end{pmatrix} \begin{pmatrix} \phi_+ \\ \phi_- \end{pmatrix} \quad \textbf{(9.35)}$$

The equations for the individual two-spinors are now decoupled, and we see that the "top" and "bottom" two-spinors are individually eigenfunctions of the energy with eigenvalues equal to plus and minus the rest energy, respectively. The appearance of this negative energy eigenvalue is completely consistent with relativity.

It should be strongly emphasized that this separation in the rest frame of reference is *not* an invariant statement; in general, the "bottom" two-spinor does not represent anti-matter as such. The entire four-component spinor represents an irreducible whole; in general, the states will have an admixture of positive *and* negative energy components. If we couple the Dirac equation to an electromagnetic field, as in the Pauli theory, then the positive and negative energy parts will be mixed together, even if they are originally decoupled. Dirac's main problem was to find a consistent interpretation of this mixing. In the next section, we shall discuss some of the real life applications of quantum mechanics to understand the operation of instruments and devices manufactured on the basis of principles and ideas developed in Sections 9.1, 9.2 and 9.3.

9.5 Real Life Applications of Quantum Mechanics

Now, we will illustrate the success of quantum mechanics in describing the behaviors that were not explained by classical mechanics. Specifically, we will discuss the applications, such as alpha particle decay, nuclear fusion in the sun, the scanning tunneling microscope, and the resonant tunneling devices fabricated using nano technology. We covered in some depth, examples of the alpha particle decay and nuclear fusion occurring inside the sun in Section 9.2.

One of the major applications of quantum mechanics is in the design of the scanning tunneling microscope (STM). The essential components of the STM are illustrated in Figure 9.6. In such a microscope, a scan of the tip over the sample can reveal the surface contours down to the atomic level. The STM enables scientists to obtain highly detailed images of surfaces at resolutions as small as the size of a single atom, which is about **0.1nm**, as indicated in Figure 8.4. For the conventional optical microscope, the resolution is limited by the wavelength of the visible light utilized to make the image, which is no more than **200nm**, that is, half the smallest wavelength of visible light. Therefore, optical microscopes are useless to observe the details on the order of the size of an atom.

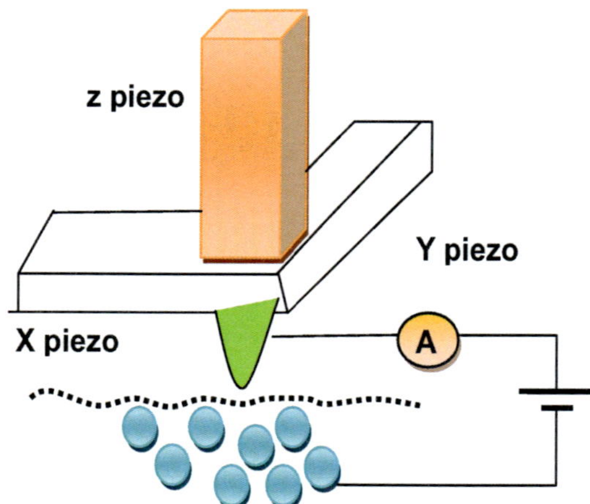

Figure 9.6 Basic arrangement of Scanning Tunneling Microscope. The electric circuit is completed by the conduction of the current through electrons and the probe tip. The tip follows the surface contour. (Courtesy of Serway & Jewett, Physics for Scientist and Engineers).

The basic idea of an STM is that an electrical voltage is applied between the surface of interest and a sharp tip. As the tip is scanned over the surface, the current varies according to the distribution of electrons in the immediate vicinity of the surface. The circuit and current path is completed by the electrons tunneled through the space (barrier) between the probe and the surface and, therefore, it is proportional to the electron count. Even though in principle, the STM looks like a simple instrument in operation, it suffers from serious drawbacks. Its operation depends on the electrical conductivity of the sample and the tip. The conductivity of most metal is poor and is affected by oxidation of the surface metals. Secondly, it is difficult to fabricate the fine tips of dimensions on the order of atom size and to prevent the oxidation on the sharpest point on the edge of the tip. A newer microscope, the atomic force microscope (AFM), overcomes these drawbacks because its operation relies on Hooke's law. However, the resolution of the AFM is limited to the size of the molecules on the order of a few hundreds of nanometers down to a few tens of nanometers. Therefore, the AFM is most commonly applied in order to perform the measurements of the molecular topologies in cells for biochemical studies. Because of the conductivity disadvantage associated with the STM, they are usually applied in the study of semiconductor material properties. In the next section, we will discuss the future of quantum physics and the role it plays in providing solutions to problems that was not solved before, by the application of quantum mechanics and relativistic physics.

9.6 Future of Quantum Physics

As we have seen, when one bridges classical mechanics with quantum mechanics, almost all of the phenomenon occurring in nature may be explained in great detail. However, a few events and facts related to fundamental particles, such as quarks and nucleons, is not explained by the unification. This opens the door for further studies and experiments to be performed by quantum physicists in the near future in order to solve the mysterious challenges. In the next paragraph, we will describe one such instance to supplement our statements and to satisfy your curiosity. One of the inadequacies of classical mechanics is that it does not explain why the mass of bonded nucleons, such as protons and neutrons in the nucleus of an atom, has lower total mass than the sum total mass of the unbound (isolated nucleon) protons and neutrons. For instance, the total mass of the helium nucleus ($^{4}\mathbf{He}$ atom, with atomic structure two protons and two neutrons) is smaller than mass of four ($^{1}\mathbf{H}$ atom, with atomic structure of one proton) hydrogen atoms. Though, the mass of two free protons and two free neutrons added together is higher than the mass of $\mathbf{4^{1}H}$ atoms. The stated discrepancy cannot be explained by the quark extension to the standard model. We saw that fact in Section 8.5 when we studied the process of the thermonuclear fusion of hydrogen into helium occurring in the sun.

Relativity, from Einstein, provided an answer to the problem, with partial satisfaction. In the theory of relativity, the total mass of the particle in motion is described by its relativistic mass, which consists of the mass at rest and the energy mass equivalent to kinetic energy possessed by the particle. Therefore, the mass of the bonded nucleon is smaller than the unbound particle because a bonded nucleon has almost zero, or negligible, kinetic mass. Hence, in a fusion type of event, the energy is released by the process, whereas in fission (opposite) kind of event, the energy needs to be supplied from the external source in order for it to occur. The deficiency of the theory of relativity is that it does not provide any account and relationship of the energy mass to the rest mass, and if it has any effect on the center of the mass of the particle. It does not state how the energy mass affects the mass distribution of the particle when kinetic energy varies. Also, it cannot explain why the total mass of the particle is not altered by the associated gravitational potential energy. The problem becomes worse by Heisenberg's uncertainty principle because it makes it much harder to measure the precise mass at rest for an unbound system of a nucleon. It is certain that unbound nucleon may not be found in the absolute rest state. We have proposed here that the mass of any particle or object has two components (refer to Section 4.4), real or rest mass, and imaginary or energy mass. This kind of splitting of the mass into components facilitates studying the energy distribution of the mass components and may provide a complete answer to our problem. This problem is an example that illustrates those principles of quantum mechanics that need to be expanded to accommodate special cases, such as the thermonuclear fusion process, to resolve the controversies. Technology needs to be devised to measure the rest mass of more fundamental particles to nucleons, such as quarks, through continuous innovation. This kind of technology will reveal many secrets and may lead to verify existence of strings more fundamental particle to quarks.

At this point, we shift our attention to an entirely new application of Dirac's equations, namely the analysis of the superconducting properties of certain materials. In particular, our focus is on the characterizing and modeling behavior of superconducting material at high temperatures, near room temperatures. It is well known that certain materials exhibit superconducting properties because for those materials, the quantum mechanical fluctuations (QMF) increase as the temperature decreases. Specifically, the materials formed from cuperate (oxides of copper), which is known to possess this property (see C. Varma and others [34]). Professor Chandra Varma predicted that QMF in the cuperate compound causes spontaneous traversing of the current loops going from copper atoms to oxygen, the oxygen atom to another oxygen atom, and then from oxygen back to copper, in a triangular fashion. These fluctuations are supposedly fluctuations in time. As per the description of Dirac's equation and Pauli's theory in Sections 9.3 and 9.4, one could understand the physical behavior and chemical properties of all the elements and chemicals in the periodic table with reasonable granularity.

We are suggesting that a set of Dirac's equations, or modified equations thereof, may be written to analyze the superconductive behavior of many materials. One of the potential applications of these equations would be to characterize the super conduction phenomena at different temperatures for existing superconducting structures. Subsequently, the analysis and techniques developed may be applied to discover and identify those materials that will show super conduction on a routine basis at high (room-ambient) temperatures. Since the advent of high temperature superconducting materials in 1987, a significant amount of progress is reported in the recent decade. The discovery and fabrication of carbon nano-tubes, and the advanced technology development, have generated a lot of interest. It is found that carbon nano tubes exhibit superconductive properties in a very high range of temperature, from **0.4K-750°K** [35]. Since the discussion about superconductivity is beyond the scope of this book, we will conclude this topic.

9.7 Summary

In this chapter, our goal was to establish a solid connection between the principles laid out by Newton for predicting the future behavior of large celestial objects, using current parameters, location, velocity, etc., with the principles laid out by Schrodinger, and others, for predicting the future behavior of microscopic particles, electrons, protons, and quarks, in the nucleus of atoms using the probabilistic wavefunction for finding the particle. We successfully developed a linkage between Newtonian mechanics and quantum mechanics, by providing the detailed facts about the operation of events of the alpha particle emission and the decay of the helium nucleus, and the thermonuclear fusion of hydrogen into helium atoms, an abundant process occurring in the sun. In Section 9.2, we analyzed the model for black body radiation in greater detail. Based on the analysis, we concluded that the quantum model for the light and oscillator energies do not invalidate classical mechanics. We state this because the frequency of radiation from any oscillator is always quantized to satisfy Pauli's exclusion principle for the quantum states of the energy radiating particle. It is Pauli's exclusion principle that places a restriction on the wavefunction for the frequency of energy radiated, such that the energy is quantized to the wave number from the particle.

In Section 9.3, we described Dirac's equation and the role he played in the formation of quantum electro dynamics. Dirac's equation essentially predicted the existence of matter and anti-matter particles with the trio of spin moment values **-1**, **0** and **+1**. From detailed understanding of Dirac's equation and theory it is clear that his equations can throw light on many theories. However his equations suffer from a limitation. In his equations it is very important to identify the observables and then treat them very carefully. The solution of his energy equations always arrive in pairs. Therefore one is tempted to predict existence of a complementary entity while he is attempting to interpret the solutions. The physicist gets trapped in search of solution to the problem on hand. In Section 9.4, we explained the issues that need to be resolved in order to connect classical mechanics with quantum mechanics. In Section 9.5, we explained the principles and operation of the scanning tunneling microscope (STM), a real life application of tunneling. This instrument is found to be highly suitable to capture detailed photographs of surfaces down to the atomic scale of **0.1nm** on semiconductor materials, with very high precision. In Section 9.6, we discussed the future of quantum physics by describing the applications, such as the thermonuclear fusion process and the analysis of the super conduction phenomena in certain compounds at high temperatures. Our explanation and analysis allowed us to conclude that the quantum ideas and techniques can be expanded to provide answers for many complex questions with total satisfaction.

A
APPENDIX: Electromagnetic Frequency Spectrum

Table A. Entire Electromagnetic Frequency Spectrum

Electromagnetic radiation is comprised of electric and magnetic fields that move at right angles to each other at the speed of light. Common examples:

Note: numbers in parenthesis are ITU bands.

Type	Band		Frequency	Wave Length [1]
	RF - Radio Frequencies		3 HZ-100 GHz	
ELF (1) - SLF (2) 1	Metal Detectors, Submarine communications		3 - 300 Hz	
ULF/VF (3)	Audio - Telephone		300 Hz-3 KHz	
VLF (4)	Navigation, Sonar,		3-30 KHz	
LF (5)	Radio navigation Beacons and Maritime		30-190 KHz	1-6 miles
MF (6)	LW	NAVTEX, Avalanche Beacon	400-500 KHz	660-750 m
	MW	BCB US AM radio	500 KHz - 1 MHz	600-175 m
	SW (4)	Ham Radio[3]	1.8-2 MHz	160 m
		Marine, Direction Finding	2-2.18 MHz	
HF (7)		Ham Radio	3.6, ... 30 MHz	80, 40, 30, 20, 17, 15, 12, 10 m
		BCB Intl. Short Wave	3.2, 4.7, 5.9, 7.1, 9.4, ... MHz	90,75,60, 49, 41, 31, 25, 22, 19, 16, 15, 13, 11 m
		CB	27 MHz	11 m
VHF (8)	Alarm Systems, Door Openers, Baby Monitors, Cordless phones		40-50 MHz	6-7 m
	VHF TV, FM Radio		54-216 MHz	5-1 m
	1¼, 2, 6 m Ham		50, 144, 222 MHz	6-1 m
	Air Traffic Control		138 MHz	
	Police, Fire		154-156 MHz	
UHF (9)	P band Radar		300 MHz	
	Remote Control (Thermometer, ..)		433 MHz	
	FRS, GMRS Walkie Talkie, Auto Racing		465 MHz	64 cm
	UHF TV		470-806 MHz	60 - 40 cm
	70, 33, 23, 13 cm Ham		420 MHz, ... 2.4 GHz	13-70 cm
	Cellular Mobil Telephone, Police		800-890 MHz	
	Cordless Phones, GSM Phones Europe		900 MHz	33 cm

		ARNS Navigation	960-1220 MHz	
	Microwaves (2)	PCS Mobil Telephone	1.8-2 GHz	15 cm
		Wi-FI, Cordless phones, Bluetooth	2.4 GHz	10 cm (4 in.)
SHF (10)		Cordless phones	5.8 GHz	5 cm
		C-band Satellite	4, 6 GHz	
		Navigation Radar	8, 12 GHz	
		Police Radar	10, 24 GHz	2-3 cm
		Ku-band Satellite	11 GHz	
		9, 5, 3, 1.2 cm Ham	3.4, 5.7, ... 24 GHz	1-9 cm
EHF (11)		LMDS	27-29 GHz	
		Radio Astronomy	31, 36, 42, 49, 98, 140, 142-149 GHz	
		Vehicle Radar	46, 76	
		6, 4, 2.5, 2, 1 mm Ham	47, 75, ...250 GHz	1-6 mm
Beyond RF			100 GHz +	
Infrared			100 GHz - 500 THz	3 mm - 700 nm
visible light waves			500 THz - 900 THz	700 - 400 nm
Ultraviolet			1 - 100 PHz	
X-rays			10 PHz - 1,000 EHz	0.3 - 3 nm
Gamma			> 1 EHz	

1. See Electromagnetic Radiation Principles for more information.

2. The boundaries between far infrared light, microwaves, and ultra-high-frequency radio waves are fairly arbitrary and are used variously between different fields of study.

3. Ham Radio - Amateur Radio

4. Shortwave - Many international stations (e.g. BBC) broadcast in English on these frequencies.

LW - Long Wave ..150-515 kHz
MW - Medium Wave..525-1,711 kHz (AM broadcast radio)
SW - Short Wave .1.711-29.9 MHz

Frequency to Wavelength Conversion Chart and Calculator

Equation: $f * \lambda = c$
Where:
f = frequency in hertz (Hz = cycles/sec)
λ = wavelength in meters (m)
c = the speed of light and is approximately equal to $2.998*10^8$ m/s

(Note: Sound waves are a form of wave energy but they are NOT electromagnetic; they are the result of mechanical vibrations creating compression/pressure waves which require a medium such as air or water to be transmitted. Electromagnetic energy can move through a vacuum like space.)

B
APPENDIX: The Greek Alphabet & English Equivalent

Greek Alphabet				English Alphabet	
Lower Case	**Upper Case**	**English Name**	**English Pronunciation**	**Lower Case**	**Upper Case**
α	Α	alpha		a	A
β	Β	beta	BAY-ta	b	B
γ	Γ	gamma		g	G
δ	Δ	delta		d	D
ε	Ε	epsilon		e	E
ζ	Ζ	zeta	ZAY-ta	z	Z
η	Η	eta	AY-ta	h	H
θ	Θ	theta	THAY-ta	q	Q
ι	Ι	iota		i	I
κ	Κ	kappa		k	K
λ	Λ	lambda		l	L
μ	Μ	mu	MEW	m	M
ν	Ν	nu		n	N
ξ	Ξ	xi		x	X
ο	Ο	omicron		o	O
π	Π	pi		p	P
ρ	Ρ	rho	ROE	r	R
σ	Σ	sigma		s	S
τ	Τ	tau	TAOW	t	T
υ	Υ	upsilon		u	U
φ	Φ	phi		f	F
χ	Χ	chi		c	C
ψ	Ψ	psi		y	Y
ω	Ω	omega	o-MEG-a	w	W
φ	ϑ			j	J
ϖ	ς			v	V

C
APPENDIX: Useful Numbers and Formulae

C.1. Physical constants

The speed of light (exact)	$c == 2.99792458\times10^{8}$ ms^{-1}
Gravitational constant	$G ==6.6726 +/- 0.0009\times10^{-11}$ m^{3} kg^{-1}s^{-2}
Planck's constant	$h == 1.054572\times10^{-34}$ Js
Dirac's constant	$\hbar = h/(2\pi) = 0.167788\times10^{-34}$ Js
Boltzmann's constant	$k == 1.3806\times10^{-23}$ JK^{-1}
Permeability of vacuum (exact)	$\mu_0 == 4\pi\times10^{-7}$ NA^{-2}
Permittivity of vacuum (exact)	$\varepsilon_0 == (c^2\mu_0)^{-1}$ Fm^{-1}
	$== 8.854187817\times10^{-12}$ C^2/Nm^2
Electron mass	$m_e == 9.109389\times10^{-31}$ kg
Proton mass	$m_p == 1.672623\times10^{-27}$ kg
Neutron mass	$m_n == 1.674929\times10^{-27}$ kg
Electron charge	$e == -1.602177\times10^{-19}$ C
Fine-structure constant	$\alpha == e^2/ (4\pi\varepsilon_0 hc) == 1/137.03599$
Thomson cross-section	$\sigma_T == e^4/(6\pi\varepsilon_0{}^2 m_e{}^2 c^4)$
	$== 6.65246\times10^{-29}$ m^2

C.2. cgs unit conversions

Energy	1 erg $== 10^{-7}$ W (watts)
Magnetic flux density	1 gauss $== 10\text{-}^4$ T (Tesla)
Charge	1 esu $== 0.1$ (c/ms^{-1})$^{-1}$ C

C.3. Astronomical units

Parsec	1 pc $== 3.0856\times10^{13}$ km
Year (sidereal)	1 yr $== 3.155815\times10^{7}$ s
Solar mass	1 $M_O == 1.989\times10^{30}$ kg
Angstrom	1 $A^0 == 1\times10^{-10}$ m $== 0.1$ nm
Jansky	1Jy $== 10^{-26}$ W m^{-2} Hz^{-1}

C.4. Cosmological quantities

Hubble parameter (dimensionless)	h	$== H_0/100\ \text{kms}^{-1}\text{Mpc}^{-1}$
Hubble time	H_0^{-1}	$== 9.7776\times10^{9}\ h^{-1}\ \text{yr}$
Hubble radius	c/H_0	$== 2997.9\ h^{-1}\ \text{Mpc}$
Density of universe	ρ	$== 1.8791\times10^{-26}\ \Omega h^2\ \text{kgm}^{-3}$
		$==2.7755\times10^{11}\ \Omega h^2\ M_O\text{Mpc}^{-3}$
Relativistic density	Ωr	$== 1.681\Omega_\gamma == 4.183\times10^{-5}\ h^{-2}$
Hydrogen density	n_H	$== 8.42\ \Omega_B h^2\ \text{m}^{-3}$
Electron density	n_e	$== 9.83\ \Omega_B h^2\ \text{m}^{-3}$
Mass per particle	μ	$== 0.593\ m_p$
Mass per electron	μ_e	$== 1.143\ m_p$

C.5. Natural and Planck units

Energy	$1\ \text{GeV} == 1.6022\times10^{-10}\ \text{J}$
Temperature	$1\ \text{GeV} == 1.1605\times10^{13}\ \text{K}$
Mass	$1\ \text{GeV} == 1.7827\times10^{-27}\ \text{kg}$
Length	$1\ \text{GeV}^{-1}== 1.9733\times10^{-16}\ \text{m}$
Time	$1\ \text{GeV}^{-1}== 6.6522\times10^{-25}\ \text{s}$
Planck mass	$m_p == (\hbar c/G)^{1/2} == 2.1767\times10^{-8}\ \text{kg}$
Planck length	$l_p == (\hbar G/c^3)^{1/2} = 1.6161\times10^{-35}\ \text{m}$
Planck time	$t_p == (\hbar G/c^5)^{1/2} = 5.3906\times10^{-44}\ \text{s}$

C.6. Conversions

Length

1 in = 2.54 cm
1m = 39.37 in = 3.281 ft
1ft = 0.3048 m

12 in = 1 ft
3 ft = 1 yd
1 yd = 0.9144 m
1 km = 0.621 mi
1 mi = 1.609 km
1 mi = 5280 ft
1 μm = 10^{-6} m = 10^{3} nm
1 light year = 9.461×10^{13} m

Area

1 m^2 = 10^4 cm^2 =10.76 ft^2
1 ft^2 = 0.0929 m^2 = 144 in^2
1 in^2 = 6.452 cm^2

Volume

1m^3 = 10^6 cm^3 = 6.102×10^4 in^3
1 lit= 10^3 cm^3 = 1.0576 qt = 0.0353 ft^3
1 ft3 = 7.481 gal = 28.32 lit
= 2.832×10^{-2} m^3
1 gal = 3.786 lit = 231 in^3

Mass
1000kg = 1 t (metric ton)
1slug = 14.59 kg
1amu = 1.66×10^{-27} kg = 931.5 MeV/c^2

Force
1N = 0.2248 lb
1lb = 4.448 N

Energy
1J = 0.738 ft. lb
1cal = 4.186 J
1 Btu = 252 cal = 1.054×10^3 J

Pressure
1bar = 14.50 lb/in^2

1atm = 14.7 lb/in2 = 1.013×10^5 N/m^2
1atm = 760 mm Hg = 76 cm Hg
1 Pa = 1 N/m^2 = 1.45×10^{-4} lb/in^2

Velocity
1 mi/h = 1.47ft/s = 0.447m/s = 1.61km/h
1 m/s = 100 cm/s = 3.281 ft/s
1 mi/min= 60 mph = 88 fps

Acceleration
1 m/s^2 = 3.28 ft/s^2 = 100 cm/s^2
1 ft/s^2 = 0.3048 m/s^2 = 30.48 cm/s^2

Power
1 hp = 550 ft.lb/s = 0.746 kW
1 W = 1 J/s = 0.738 ft.lb/s^2
1 Btu/h = 0.293 W
1 kWh = 3.60×10^6 J

Time
1 Yr = 365 days = 3.16×10^7s
= = 31557600 s
1 day = 24 hr = 1.44×10^3 min
= 8.64×10^4 s

D
APPENDIX: Glossary

Accelerator: A machine used to accelerate particles to high the speeds (and thus high energy, as compared to their rest mass energy)

Angular momentum: Angular momentum is a conserved quantity, which is used to describe rotational motion much like momentum for linear motion. Rotational motion can be orbital motion of two bodies around one another or the rotation of a rigid body. The intrinsic angular momentum of a particle is called "spin." In quantum mechanics, angular momentum and spin are quantized quantities: They can only have certain discrete values, measured in multiples of $\hbar$, which is Planck's constant h divided by 2π.

Annihilation: A process in which a particle meets its corresponding antiparticle and both disappear. The energy appears in some other form, perhaps as a different particle and its antiparticle (and their energies), perhaps as many mesons, and perhaps as a single, neutral boson, such as a Z^0 boson. The produced particles may be any combination allowed by conservation of energy, momentum, electric, magnetic and other charge types and by other rules.

Antifermion: The antiparticle of a fermion. See also antiparticle.

Antimatter: Material made from antifermions. We define the fermions that are common in our universe as matter and their antiparticles as antimatter. In the particle theory, no a priori distinction exists between matter and antimatter. The asymmetry of the universe between these two classes of particles is a deep puzzle for which we are not yet completely sure of explanation.

Antiparticle: For every fermion type, another mirror fermion type exists that has exactly the same mass but the opposite value of all other charges (quantum numbers). This is called the antiparticle. For example, the antiparticle of an electron is a particle of positive electric charge called the positron. Bosons also have antiparticles, except for those that have zero value for all charges, for example, a photon or a composite boson made from a quark and its corresponding antiquark. In this case, the particle and the antiparticle have no differences; they are the same object.

Antiquark: The antiparticle of a quark. An antiquark is denoted by putting a bar or apostrophe over the corresponding quark symbol (d', u', s' etc.).

Atom: Fundamental building block of matter, consisting of a nucleus (comprising protons and neutrons) and an orbiting swarm of electrons.

Astrophysics: The physics that studies astronomical objects such as stars and galaxies.

AU: Unit of distance which corresponds to average orbit distance of planet earth from the Sun. 1AU = 1.496×10^8 km = 9.2977×10^6 miles

B Factory: An accelerator designed to maximize the production of B mesons. The properties of B mesons are then studied with specialized detectors.

Baryon [BARE-ee-on]: A hadron made from three quarks. The proton (uud) and the neutron (udd) are both baryons. They may contain additional quark-antiquark pairs.

Baryon-antibaryon asymmetry: The observation that the universe contains many baryons but few antibaryons: a fact that needs explanation.

Beam: The particle stream produced by an accelerator usually clustered in bunches.

Big Bang Theory: The theory of an expanding universe that begins as an infinitely dense and hot medium. The initial instant is called the Big Bang.

Black hole: An object whose immense gravitational field (belief) entraps anything, even light, that gets too close (closer than the black hole's event horizon).

Boson [BOZE-on]: A particle that has integer intrinsic angular momentum (spin) measured in units of h' (spin = 0,1,2…..). All particles are either fermions or bosons. The particles associated with all fundamental interactions (forces) are bosons. Composite particles with even number of fermion constituents (quarks) are also bosons.

Bottom quark (b): The fifth flavor of quark (in order of increasing mass), carrying electric charge -1/3

Brane: Any of the extended objects that arise in string theory. A one-brane is a string, a two-brane is a membrane, a three brane has three extended dimensions, etc. More generally, a p-brane has p spatial dimensions.

Calorimeter: A device that can measure the energy deposited in it (originally, devices to measure thermal energy deposited, using change of temperature; particle physicists use the word for any energy-measuring device).

CERN: The major European international accelerator laboratory **Center for Electronics Research in Nuclear Physics** which is located near Geneva, Switzerland.

Charge: A quantum number carried by a particle. This quantum number determines whether the particle can participate in an interaction process. A particle with electrical charge has electrical interactions; one with strong charge has strong interactions, etc.

Charge conservation: The observation that electric charge is conserved in any process of transformation of one group of particle into another.

Charm quark (c): The fourth flavor of quark (in order of increasing mass), carrying electric charge +2/3

Chirality: Feature of fundamental particle physics that distinguishes left from right-handed, showing that the universe is not fully left-right symmetric. This feature occurs as a result of random law of nature and entropy of random process is always increasing.

Collider: An accelerator in which two beams travelling in opposite directions are steered together to provide high-energy collisions between the particles in one beam and those in the other.

Colliding-beam experiments: Experiments done at colliders.

Color charge: The quantum number that determines participation in strong interactions; quarks and gluons carry nonzero color charges.

Color neutral: An object with no net color charge. For composite made of color-charged particles, the rules of neutralization are complex. Three quarks (baryon) or a quark plus an antiquark (meson) can both form color-neutral combinations.

Complex mass (Skylativity® Theory): The total mass (m) of a particle is a complex phasor with a static or fix component eternal mass and imaginary orthogonal component descriptive of state of kinetic and potential energy associated with the mass. The mass distribution of eternal real part of the mass affects the position of center of gravity for the mass. Symbolically

$$\mathbf{M_t = M_r + j\ M_e}$$

The imaginary component of the mass M_e varies depending on energy and momentum configuration. (See also mass)

Confinement: The property of the strong interactions by which quarks or gluons are never found separately but only inside color-neutral composite objects.

Conservation: When a quantity (eg. Electric charge, energy, or momentum) is conserved; it is the same after a reaction between particles as it was before.

Cosmic microwave background radiation: Microwave radiation suffusing the universe, produced during the big bang and subsequently thinned and cooled as the universe expanded.

Cosmology: The study of the history of universe.

Dark matter: Matter that is in space but is not visible to us because it emits no radiation by which we can observe it. The motion of stars around the centers of their galaxies implies that about 90% of the matter in a typical galaxy is dark. Physicists speculate that dark matter also exists between the galaxies, but this is harder to verify.

Decay: A process in which a particle disappears and in its place two or more different particles appear. The sum of the masses of the produced particles is always less than the mass of the original particle. (The mass-energy is conserved, however.)

Detector: Any device used to sense the passage of a particle. Also, the word detector is used for a collection of such devices designed so that each serves a particular purpose in allowing physicists to reconstruct particle events.

Dimension: An independent axis or direction in space or spacetime. The familiar space around us has three dimensions (left-right, back-forth, up-down) and the familiar spacetime has four (the previous three axes plus the past-future axis). Superstring theory requires the universe to have additional spatial dimensions.

Dirac's Equation: Paul Dirac derives an equation that combines quantum mechanics and special relativity to describe the electron; it is found to also require the existence of corresponding positively charged particles. Dirac, together with Schrodinger, is awarded Nobel Prize in 1933.

Dirac, Paul : See Dirac's equation.

Down quark (d): The second flavor of quark (in order of increasing mass), carrying electric charge -1/3

Dynamic mass: The portion of mass of an object that represents quantity of energy contained in the object because of dynamics. This is a fictional or imaginary component of the total mass of the object. (See also eternal mass.)

E_d: Equatorial diameter of earth used to express sizes of other members of solar system. 1 E_d = 12756 km = 7927.91 miles

Electric charge: The quantum number that determines participation in electromagnetic interactions.

Electromagnetic interaction: The interaction due to electric charge; this includes magnetic effects, which have to do with moving electric charges.

Electron [e-LEC-tron] (e): The least-massive electrically charged particle, hence, absolutely stable. It is the most common lepton, with electric charge -1.

Electroweak interaction: In the standard model, electromagnetic and weak interactions are related (unified); physicists use the term electroweak to encompass both of them.

E_m: Mass of earth used to express sizes of other members of solar system.
$1\ E_m = 3.302\times10^{23}$ kg
Entropy: A measure of the disorder of a physical system, the number of rearrangements of the ingredients of a system that have its overall appearance intact.

Euclidean Topology: A geometrical surface with property that the shortest distance between two points on the surface is a straight line. For a Non-Euclidean topology surface the shortest distance is a curved line.

Eternal mass: That portion of mass of an object that determines the position of center of gravity for the object. (See also rest mass.)

Event: What occurs when two particles collide or a single particle decays? Particle theories predict the probabilities of various possible events occurring when many similar collisions or decays are studied. They cannot predict the outcome for any single event.

Event horizon: The one-way surface of a black-hole; once penetrated, the laws of gravity ensure that there is no turning back, no escaping the powerful gravitational grip of the black hole.

Exclusion principle: Wolfgang Pauli's principle which states that no two fermions can exist in the same state at the same place and time. Many of the properties of ordinary matter arise because of this rule. Electrons, protons, and neutrons are all fermions, as are all fundamental matter particles, quarks and leptons.

Fermilab: Fermi National Accelerator Laboratory in Batavia, Illinois (near Chicago). The lab oratory was named after particle physics pioneer Enrico Fermi.

Fermion [FARE-mee-on]: Any particle that has odd, half-integer (1/2, 3/2,….) intrinsic angular momentum (spin), measured in units of h. As a consequence of this peculiar angular momentum, fermions obey a rule called **the Pauli Exclusion Principle**, which states that no two fermions can exist in the same state at the same place and time. Many of the properties of ordinary matter arise because of this rule. Electrons, protons, and neutrons are all fermions, as are all fundamental matter particles, quarks and leptons.

Fixed-target experiment: An experiment in which the beam of particles from an accelerator is directed at a stationary (or nearly stationary) target. The target may be a solid, a tank containing liquid or gas, or a gas jet.

Flavor: The name used for different quark types (up, down, strange, charm, bottom, top) and for the different lepton types (electron, muon, tau). For each charged lepton flavor, a corresponding neutrino flavor exists. In other words flavor is the quantum number that distinguishes the different quark/lepton types. Each flavor of quark and charged lepton has a different mass. For neutrinos, we do not yet know if they have a mass or what the masses are.

Freeze out: As the universe expands and cools, the probability of any collision-driven process decreases, because the rate of necessary collision decreases. A process can be ignored when the average time between collisions is long compared to the age of universe at that time. Such a process is then said to have frozen out.

Fundamental interaction: In the Standard Model, the fundamental interactions are the strong, electromagnetic, weak, and gravitational interactions. At least one more fundamental interaction (Higgs) is in theory; it is responsible for fundamental particle masses. Five interactions types are all that are needed to explain all observed physical phenomena.

Fundamental particle: A particle with no internal substructure. In the Standard Model, the quarks, leptons, photons, gluons, W^+ & W^- bosons, and Z^0 bosons are fundamental. All other objects are made from these.

Galaxy: A collection of stars held together by gravitational forces.

General relativity: The theory of gravitation formulated by Einstein.

Generation: A set of one of each charge type of quark and lepton, grouped by mass. The first generation contains the up and down quarks, the electron, and the electron neutrino.

Gluon [GLUE-on] (g): The carrier particle of strong interaction.

Grand unified theory (GUT): Any of class of theories that contain the Standard Model, but go beyond it to predict further types of interactions mediated by particles with masses of order 10^{15}GeV/c^2. At large energies compared to this mass (times c^2), the strong, electromagnetic, and weak interactions are seen as only different aspects of one unified interaction.

Gravitational interaction: The interaction of particles due to mass-energy.

Graviton: The carrier particle of the gravitational interaction (not yet directly observed).

GTR: Einstein's general theory of relativity.

Hadron [HAD-ron]: A particle made of strongly interacting constituents (quarks and/or gluons). These include the mesons and baryons. Such particles participate in residual strong interactions.

Half-life: The rate of decay of a radioactive material measured by how long it would take for half of the atoms in a given bunch to have randomly decayed.

Higgs boson: The carrier particle (or quantum excitation) of the additional force needed to introduce particle masses into the Standard Model (not yet observed).

Interaction: A process in which a particle decays or annihilates or it responds to a force due to presence of another particle (as in a collision). Also used to mean, underlying property of the theory that causes such effect.

Interference pattern: Wave pattern that emerges from the overlap and the intermingling of waves emitted from different locations.

ITU: International Telecommunication Union Standards, formerly CCITT, produces global telecommunication standards and defines tariff & accounting principles.

Jet: Depending on their energy, the quarks and gluons emerging from a collision will materialize into 5-30 particles (mostly mesons and baryons). At high momentum, these particles will appear in clusters called "jets" that is, in groups of particles moving in roughly the same direction, centered about the path of original quark or gluon.

Kaon (K): A meson containing a strange quark and an anti-up (0r an anti-down) quark, or an anti-strange quark, and an up (or down) quark.

Laplacian determination: Clockwork conception of the universe in which complete knowledge of the state of the universe at one moment is completely determines its state at all feature and past moments.

Laser: Light amplification by stimulated emission of radiation (LASER or laser)

Light clock: A hypothetical clock that measures elapsed time by counting the number of round trip journeys completed by a single photon (hypothetical particle) between two mirrors.

Lorentz contraction: Feature emerging from the special relativity, in which a moving object **appears** shortened along its direction of motion.

Lepton (LEP-tahn): A fundamental fermion that does not participate in strong interactions. The electrically charged leptons are the electron (e), muon (μ), tau (τ), and their anti-particles. Electrically neutral leptons are called neutrinos (υ).

LHC: The Large Hadron Collider at the CERN laboratory in Geneva, Switzerland. LHC collide protons with protons at a center-of-mass energy of about 14 TeV. When completed in year 2005, it will be the most powerful accelerator in the world. It is hoped that it will unlock many of the remaining secrets of particle physics.

Light year: The astronomical unit of length. A light year (ly) is the distance light travels during a year; $1 \text{ ly} = 0.9461 \times 10^{13}\text{km}$ = 6 trillion miles approximately.

Life time: The time between the creation and the decay of a type of particle. The life time of an individual particle cannot be predicted. We can just measure an average (or mean) life time by observing the random decay in a sample of a given type of unstable particles.

Linac: An abbreviation for linear accelerator, that is, an accelerator that is straight.

Maser: Microwave amplification by stimulated emission of radiation (MASER or maser)

Mass (Current Theory): The mass (m) of a particle is the mass defined by the energy of the isolated (free) particle at rest, divided by c^2. (This is very slightly correct. The majority of particle mass consists of protons and neutrons. Baryon rules states number of protons and neutrons is unaltered in any mass to energy conversion process.) When particle physicists use the word "mass" they always mean the "rest mass" (m) of the object in question. The total energy of a free particle is given by

$$\mathbf{E = (p^2c^2 + m^2c^4)^{1/2}}$$

Where p is the momentum of the particle. Note that for p = 0 this simplifies to Einstein's $\mathbf{E = mc^2}$. For a general particle with mass and momentum it can also be written as $\mathbf{E == \gamma mc^2}$, where $\boldsymbol{\gamma = (1\text{-}v^2/c^2)^{-1/2}}$. Some textbooks on special relativity identify γm as the "mass" of a moving particle; this definition is not used in particle physics. The quantity E includes both the mass-energy and the kinetic energy. (This definition is biased to kinetic energy it does not take into account changes in potential energy of particles. Energy conservation rule applies to potential and kinetic energy both).

Maxwell's electromagnetic theory: Theory uniting electricity and magnetism, based on the concept of the electromagnetic field, devised by Maxwell in the 1880's; shows that visible light is an example of an electromagnetic wave.

Meson [MEZ-on]: A hadron made from an even number of quark constituents. The basic structure of most mesons is one quark and one anti-quark; some of multiples of this.

Microwave: An electromagnetic wave with wavelength in the micrometer range.

Muon [MEW-on] (μ): The second flavor of charged lepton (in order of increasing mass), with electric charge -1.

Muon chamber: The outer layers of a particle detector capable of registering tracks of charged particles. Except for the charge-less neutrinos, only muons reach this layer from the collision point.

Neutral: Having a net charge equal to zero. Unless specified otherwise, it usually refers to electric charge.

Neutrino [new-TREE-no] (υ): A lepton with no electric charge. Neutrinos participate only in weak and gravitational interactions and, therefore, are very difficult to detect. Three known types of neutrino exist, all of which are very light and could possibly even have zero mass.

Neutron [new-TRON] (n): A baryon with electric charge zero; it is a fermion with a basic structure of two down quarks and one up quark (held together by gluons). The neutral component of an atomic nucleus is made from neutrons. Different isotopes of the same element are distinguished by having different numbers of neutrons in their nucleus.

Newton's universal theory of gravity: Theory of gravity declaring that the force of attraction between two bodies is proportional to the product of their masses and inversely proportional to the square of the distance between them.

NP incomplete: A problem whose solution requires infinite computational resource for real time computation of results. NP complete problems may be solved with finite resource in real time.

Nucleon: A proton or a neutron; that is, one of the particles that makes up a nucleus.

Nucleosynthesis: The process by which protons and neutrons combined to form nuclei in the early universe.

Nucleus: A collection of neutrons and protons that forms the core of an atom (plural: nuclei).

Parsec (pc): An astronomical unit of length. It is equal to the distance at which the sun-earth separation subtends an angle of one second; 1 pc = 3.2616 light years.

Particle: A subatomic object with a definite mass and charge.

Pauli, Wolfgang: See Exclusion principle.

Perturbation theory: Framework for simplifying a difficult problem by finding an approximate solution that is subsequently refined as more details initially ignored are systematically included.

Photon [FOE-than] (γ): The carrier particle of electromagnetic interaction.

Phase: When used in reference to matter, describes its possible states, solid phase, liquid phase, gas phase. More generally refers to the possible descriptions of a physical system as features on which it depends (temperature, string coupling constant values, form of spacetime, etc.) are varied.

Phase transition: Evolution of a physical system from one phase to another.

Photoelectric effect: Phenomenon in which electron are ejected from a metallic surface when the surface is exposed to light.

Pion [PIE-on] (π): The least massive type of meson, pion can have electric charges +/- 1or 0

Planck energy: About 1,000 Kilowatt hours, the energy necessary for probing to distances as small as the Planck length. This is the amount of typical energy of a vibrating string in string theory.

Planck length: About 10^{-33} centimeters, the scale below which quantum fluctuations in the fabric of spacetime would become enormous. This is the size of a typical string in string theory.

Planck mass: About 10^{18} times the mass of a proton; about 10^{-5} times the mass of a gram; about the mass of a small grain of dust. This is the typical mass equivalent of a vibrating string in string theory.

Planck constant: Denoted by the symbol h', Planck's constant is a fundamental parameter in quantum mechanics. It determines the size of the discrete units of energy, mass, spin, etc. into which the microscopic world is partitioned. Its value is 1.05×10^{-27} grams-cm/s. This constant characterizes a basic particle into integral multiples of parameters mass, energy etc.

Planck tension: About 10^{39} tons. This is the tension on a typical string in string theory.

Planck time: About 1,000 Kilowatt hours, the energy necessary for probing to distances as small as the Planck length. This is the amount of typical energy of a vibrating string in string theory.

Plasma: A gas of charged particles.

Positron [PAUSE-i-tron] (e^+): The antiparticle of the electron. In 1931, Paul Dirac realizes that the positively charged particles required by his equation are new objects (he calls them "positrons"). They are exactly like electrons; in particular, they have the identical mass, but positively charged. The positrons are first example of antiparticles.

Principle of equivalence: Core principle of general relativity declaring the indistinguishability of accelerated motion immersion in a gravitational field (over small enough regions of observation) Generalizes the principle of relativity by showing that all observers, regardless of their state of motion, can claim to be at rest, so long as they acknowledge the presence of a suitable gravitational field.

Principle of relativity: Core principle of special relativity declaring that all constant velocity observers are subject to an identical set of physical laws, and that therefore, every constant velocity observer is justified in claiming that he or she is at rest. This principle is generalized by the principle of equivalence.

Proton [PRO-than] (p): The most common hadron, a baryon with electric charge (+1) equal and opposite to that of electron (-1). Proton has a basic structure of two up quarks and one down quark (bound together by gluons). The nucleus of a hydrogen atom is a proton. A nucleus with electric charge Z contains Z protons; therefore, the number of proton is what distinguishes the different chemical elements.

Quantum: The smallest discrete amount of any quantity (plural; quanta).

Quantum chromodynamics (QCD): Relativistic quantum field theory of the strong force and quarks, incorporating special relativity.

Quantum determinism: Property of quantum mechanics that knowledge of the quantum state of a system at one moment completely determines its quantum state at future and past moments. Knowledge of the quantum state, however, determines only the probability that one or another future will actually ensue.

Quantum electrodynamics (QED): Relativistic quantum field theory of the electromagnetic force and electrons, incorporating special relativity.

Quantum geometry: Modification of Riemannian geometry required describing accurately the physics of space on ultramicroscopic scales, where quantum effects become important.

Quantum gravity: A theory that successfully mergers quantum mechanics and general relativity, possibly involving modifications of one or both. String theory is an example of a theory of quantum gravity.

Quantum mechanics: The law of physics that apply on very small scales. The essential feature is that energy, momentum, and angular momentum, as well as charges comes in discrete amounts called quanta.

Quark [KWORK] (q): A fundamental fermion that has strong interactions. Quarks have electric charge of either 2/3 (up, charm, top) or -1/3 (down, strange, bottom) in units where the proton charge is 1

.**Residual interaction:** Interaction between objects that do not carry a charge but do contain constituents that have charge. Except for those chemical substances involving electrically charged ions, much of chemistry is due to residual electromagnetic interactions between electrically neutral atoms. The residual strong interaction between protons and neutrons, due to the strong charges of their quark constituents, is responsible for the binding of the nucleus.

Rest mass: See eternal mass.

Riemannian geometry: Mathematical framework for describing curved shape of any dimension. The geometry plays a central role Einstein's description of spacetime in general relativity.

Schrödinger equation: Equation governing the evolution of probability waves in quantum mechanics.

Schwarzschild solution: Solution to equation of general relativity for a spherical distribution of matter: one implication of this solution is the possible existence of black holes.

Scintillation: A charged particle traversing matter leaves behind it a wake of excited molecules. Certain types of molecules will release a small fraction of this energy as light. This light can be detected by a phototube in a detector.

Second law of thermodynamics: Law stating that total entropy always increases.

Skylativity® Theory: New theory of relativity developed by Shailesh Kadakia in 2005 for applications to Sky and Stellar celestial objects.

SLAC: The Stanford Linear Accelerator Center in Stanford, California

Spin: Intrinsic angular momentum of a particle, given in units of h', the quantum unit of angular momentum,

where $\mathbf{\hbar == h/2\pi == 6.58\times10^{-34}}$ **J s.** Spin is a characteristic property for each type of particle.

Stable: Does not decay. A particle is stable if no processes exist in which the particle disappears and in its place two or more different particles appear.

Standard Model: Physicists' name for the theory of fundamental particles and their interactions, as described in this book. It is widely tested and accepted as correct by particle physicists.

STR: Einstein's special theory of relativity.

Strange quark (s): The third flavor of quark (in order of increasing mass), carrying electric charge -1/3.

String coupling constant: A (positive) number that governs how likely it is for a given string to split apart into two strings or for two strings to join together into one—the basic processes in string theory. Each string theory has its own string coupling constant, the value of which should be determined by an equation; currently such equations are not understood well enough to yield any useful information. Coupling constant less than 1 imply that perturbative methods are valid.

String theory: Unified theory of the universe postulating that fundamental ingredients of nature are not zero-dimensional point particles but tiny one-dimensional filaments called strings. String theory harmoniously unites quantum mechanics and general relativity, the previously known laws of the small and the large, which are otherwise incompatible. It is often used a short form for superstring theory.

Strong interaction: The interaction responsible for binding quarks, antiquarks, and gluons to make hadrons. Residual strong interactions provides the nuclear binding force.

Subatomic particle: Any particle that is small compared to the size of the atom.

Supernova: An old star that has burnt most of its hydrogen collapses due to gravitational attraction, but then explodes from the onset of nuclear burning of more massive elements.

Synchrotron: A type of circular accelerator in which the particles travel in synchronized bunches at fixed radius.

Tachyon: Particle whose mass (squared) is negative; its presence in a theory generally yields inconsistencies. For instance tachyons are believed to have the speed above the speed of light.

Tau [TAOW] lepton: The third flavor of charged lepton (in order of increasing mass), with electric charge -1.

Tevatron Collider: An accelerator at Fermilab that collides protons and antiprotons with center-of-mass energy of 2 TeV (2000 GeV).

Theory of everything: A quantum-mechanical theory that encompasses all forces and all matters.

Time dilation: Feature emerging from special relativity, in which flow of time apparently slows down for an observer in motion.

Top quark: The sixth flavor of quark (in order of increasing mass), with electric charge 2/3. Its mass is much greater than any other quark or lepton.

Track: The record of the path of a particle traversing a detector.

Tracking: The reconstruction of a "track" left in a detector by the passage of a particle through the detector.

Uncertainty Principle: The quantum principle, first formulated by Heisenberg, that states that it is not possible to know exactly both the position x and its momentum p of an object at the same time, $\Delta x \Delta p \geq \frac{1}{2} \hbar$. It can be written as $\Delta E \, \Delta t \geq \frac{1}{2} \hbar$ where ΔE means the uncertainty in energy and Δt means the uncertainty in life time of a state (see virtual particle).

Such uncertain aspects of the microscopic world become even more severe as the distance and time scales on which they are considered become ever smaller. Particles and fields undulate and jump between all possible values consistent with the quantum uncertainty.

Up quark: The least massive flavor of quark with electric charge 2/3.

Vertex detector: A detector placed very closed to the collision point in a colliding-beam experiment so that tracks coming from the decay of a short-lived particle produced in the collision can be accurately reconstructed and seen to emerge from a "vertex" point that is different from the collision point.

Virtual particle: A particle that exists only for an extremely brief instant as an intermediary in a process. The intermediate or virtual particle stages of a process cannot be directly observed. If they were observed, we might think that conservation of energy was violated. However, the Heisenberg Uncertainty Principle (which can be written as $\mathbf{\Delta E\ \Delta t \geq ½\ h'}$) allows an apparent violation of the conservation of energy. If one sees only the initial decaying particle (such as meson with the c quark) and the final decay products (such as $\mathbf{s} + \upsilon_e + \mathbf{e}^+$), one observes that energy is conserved. The "virtual" particle (such as the $W^{+/-}$) exists for such a short time that it can never be observed.

VSL: Varying the speed of light theory proposed by Portuguese physicist and Cambridge University Professor, João Magueijo, in 1995 [16].

$W^{+/-}$ boson: A carrier particle of the weak interactions. It is involved in all electric-charge-changing weak processes.

Weak interaction: The interaction responsible for all processes in which flavor changes, hence for the instability of heavy quarks and leptons, and particles that contain them. Weak interactions that do not change flavor (or charge) have also been observed.

Wormhole: A tube-like region of space connecting one region of the universe to another. In view of the fact that our universe has no boundaries (infinite) existence of wormhole does make sense. It is a fictional entity.

Z^0 boson: A carrier particle of weak interactions. It is involved in all weak processes that do not change flavor.

E
APPENDIX: History of Particle & Space Science

E.1. Early Understanding of Atomic Structure

400BC	Democritus discusses the idea that everything is made from individual (fundamental) particles, which he called "atoms"
1803	John Dalton formulates the "law of definite proportions:" That the relative amounts of the elements which are constituents in a particular chemical compound are always the same, regardless of origin or method of preparation.
1808	Joseph-Louis Proust proves experimentally Dalton's "law of definite proportions."
1815	Proust hypothesizes that the atomic weights of elements are whole number multiples of the atomic weight of hydrogen.
1859-1861	Kirchhoff and Bunsen measure wavelengths of atomic spectral lines, establishing that spectra are unique to each element. They and others use spectral analysis to identify new elements. About this time, a general suspicion among researchers develops that the spectra characterize the atomic species. Several models or pictures of the basic constituents of matter are suggested.
1861-1865	Maxwell, in a series of papers, describes the interrelation of electric and magnetic fields, thereby unifying them into electromagnetism. This leads to now-famous Maxwell's Equations. One prediction of these equations is that there are traveling electromagnetic waves in addition to static electric and magnetic fields.
1867	Kelvin proposes a vortex atom, a geometrical structure that was a stable assembly of interlocking vortex rings. Kelvin conjectures that a classification of knots would yield a classification of the elements.
1867	Discovery of unusual rays by Crookes in gas discharges. Named "cathode rays," they were believed to be a fourth state of matter; now cathode rays are understood to be a stream of electrons.
1869	Mendeleev classifies the known chemical elements in a "periodic table" according to atomic mass and chemical properties, with gaps for unknown elements.

1875	Maxwell notes that atoms have a structure that is much more complicated than that of a rigid body, that is to say, some internal motion is possible.
1878	Lorentz, in his inaugural address at the University of Leyden, discusses the structure of matter. Matter is subdivided into molecules, which are composed of atoms. Atoms are characterized by their optical properties (spectra).
1881	Helmholtz argues that the existence of chemical atoms plus Faraday's laws of electrolysis (the fact that certain chemical processes produce electric current) imply that electricity consists of finite elementary units.
1881-1884	Hertz demonstrates radio waves and establishes that both radio waves and light are electromagnetic waves of different frequencies verifying Maxwell's theory.
1895	Perrin observes that cathodes rays are negatively charged.
1896	Henri Becquerel accidently discovers that uranium emits radiation that exposes a sensitive film-plate. While investigating how fluorescent uranium exposes the plate, he recognizes that non-fluorescent uranium has the same effect, even when wrapped in black paper. He further studies and finds that the radiation could also penetrate through silver pieces. This was the first recognition of radioactivity.
1897	Marie Curie, working under very primitive laboratory conditions, succeeds in isolating two radioactive elements, polonium and radium, from pitchblende.
1897-1899	J.J. Thompson, in a series of experiments, shows that cathode rays consist of negatively charged corpuscles. He finds that the same charge to mass ration no matter how cathode rays are produced. He concludes that the corpuscles are universal constituents of all atoms and that their mass is about 1/1000 of that of a hydrogen atom. Thus he discovers the electron.
1899	It is recognized that there are three different types of rays which are named a, b, g. Rutherford recognizes that the rays are emitted in decay of radium are positively charged particles, which he named a particles. These are now known to be a helium nucleus. It begins to be clear that one atom can "transmute" through radioactive decay into a different atom, which destroys the notion that each element is an "elementary particle" in its own right, though it takes another 40 years for a clear picture of what the nucleus is to emerge.
1900-1903	J.J. Thompson, about 1900, proposes a model for the atom: a number of electrons moving in some vaguely specified, positive background of indefinite shape. As late as 1903, Thompson thought that a hydrogen atom contained

about 1000 electrons. Lorentz thought that an electron was an extended charge distribution.

E.2. Development of Quantum Ideas.

1900 Planck, trying to understand the observed spectrum of energy radiated from a blackbody, that is any black object at a temperature warmer than its surroundings, suggests that the radiation is quantized: that is, it comes in a certain discrete amounts. Planck presents his ideas to an audience of not quite 25 people. Nobody understands much, except the final formula, which fir measurements that had previously been known but unexplained. Only a few physicists pay attention to the quantum ideas; there is much confusion. No obvious connection is made in Planck's treatment between the quantum radiation and the structure of matter.

1905 Einstein, one of the few physicists then taking Planck's quantum ideas seriously, proposes that light consists of discrete energy packets (now called photons) whose energy is proportional to frequency. This provides for an explanation of the photoelectric effect, in which light falling on a surface ejects electrons from the surface. But nobody, including Planck, takes it seriously at first. (This is the work for which Einstein is awarded the Nobel Prize for 1921. According to authors of this book his work on photoelectric effect and other work, on relativity, is regarded as too speculative.

1911 Geiger and Marsden, in an experiment studying the scattering of particles from thin foils, observe that a few particles scatter at large angles, some even backwards; Rutherford realizes that this suggests the existence of a small hard core inside an atom.

1911 Rutherford proposes the nuclear model of an atom, which is violently opposed by J.J. Thompson. The reception of Rutherford's ideas is decidedly mixed.

1911 The first Solvay conference is held in 1911 in Brussels; attendance is by invitation only. (Ernest Solvay was an industrial chemist who had always aspired to be a theoretical physicist. Instead he became extremely wealthy. On the urging of some scientist (Nernst and others), he lavishly supported a series of physics conferences, dealing with fundamental current problems. Solvay conferences are still held in Brussels about once every two or three years). At this conference, quanta are taken extremely seriously. The nuclear model of an atom with a localized positive charge is never mentioned.

1913 Bohr succeeds in constructing a theory of atomic electronic structure based on quantum ideas. His theory introduces the idea of quantum states (or quantum levels) for the electrons in an atom and quantum jumps (or quantum transitions) of electrons between these levels as the mechanism that produces the characteristic spectral lines of radiation from atoms.

1919 Rutherford observes first nuclear transmutation $N_7^{14} + \alpha \rightarrow O_8^{17} + p$. This process provides the first evidence that object recognized to be the nucleus of the hydrogen atom is also a constituent of other nuclei. Two years later, Rutherford postulates that it is a fundamental particle and names it the proton. Rutherford also makes a first estimate of nuclear size.

1921 From studies on α-hydrogen scattering, Chadwick and Bieler conclude that some kind of force (not following the $1/r^2$ force law) exists inside the nucleus.

1923 Compton discovers the particle nature of X-rays, as quanta processing both energy and momentum.

1924 Louis de Broglie conjectures the wave aspect of matter. He introduces the wave-particle duality as a universal feature for all types of matters and radiation.

1925 January: Pauli formulates the exclusion principle for electrons in an atom. He recognizes the need for additional parameter (that can take two values) in the quantum properties of the electron, later seen as the two spin orientations. Nobel Prize awarded in 1945 for this work.

1925 April: First experimental demonstration of energy and momentum conservations in individual processes (by Bothe and Geiger).

1925 October: Goudsmit and Uhlenbeck introduce the spin as a new electron attribute, assigning intrinsic angular momentum of ½ h to the electron.

1925-1926 Heisenberg invents matrix mechanics and within months Schrodinger develops wave mechanics; both are formulations of quantum theories to explain atomic systems. Soon after, Dirac shows that the two theories are equivalent within more general framework. Max Born gives a probabilistic, statistical meaning to Schrödinger's wavefunction. G.N. Lewis proposes the name "photon" for a light quantum.

1927 Certain materials (e.g., radium) are observed to emit electrons, in the process known as β decay. The spectrum of electron energies is shown to be continuous (non-quantized). For a while, it is unclear whether these electrons are of nuclear or atomic origin. Since both the atom and the nucleus are thought to have discrete energy levels, it is hard to see how electrons produced in either type of transition could have a continuous spectrum.

1927 Heisenberg formulates the uncertainty relations, which state the impossibility of making simultaneous, arbitrarily precise measurements of both a particle's momentum in some direction and its coordinate in that direction. Later we shall see that his principles are a direct consequence of the Second law of Kadakia that no measurements which involve the speed faster than absolute the speed of light can be performed.

1927 Wigner introduces the concept of "parity" of quantum states (a consequence of left-right symmetry), a labeling of states by the properties (even or odd) of their wavefunction s under reflection of all directions about the origin. The fact that states have definite parity is a consequence of an invariance of the equations describing the interactions under such a change of coordinate definitions. This extra label explains Laporte's rule in atomic spectra. Parity is rapidly recognized as an essential attribute of quantum states.

1928 Dirac derives an equation that combines quantum mechanics and special relativity to describe the electron; it is found to also require the existence of corresponding positively charged particles. Dirac, together with Schrodinger, is awarded Nobel Prize in 1933.

1929 Dirac proposes (incorrectly) that the positively charged particles required by his equations are protons.

1930 A general consensus is reached. Quantum mechanics and special relativity are well established. Just three fundamental particles exists: protons, electrons and photons. Born after learning about the Dirac equation (1929), said, "Physics as we know it will be over in six months."

E.3. Transition from Atomic to Particle Physics.

1930 Pauli, in letters and private conversations, suggests that an additional new type of particle, the neutrino, must be being produced to explain the continuous electron spectrum for β decay. (See 1927. If particles are produced in the transition, only the sum of their energies is discrete.)

1931 Dirac realizes that the positively charged particles required by his equation are new objects (he calls them "positrons"). They are exactly like electrons; in particular, they have the identical mass, but positively charged. The positrons are first example of antiparticles.

1931-1932 Anderson observers positively charged particles (produced by cosmic rays) that have the same mass as electrons. Pauli and Bohr do not believe that these are Dirac's positrons, but they are later determined to be just.

1931	Discovery of the neutron, its spin is determined to be 1/2h. Chadwick awarded Nobel Prize in 1935. It is suggested that a neutron is as elementary as a proton. Nuclei can now be understood as composed of protons and neutrons. The questions of the mechanisms of nuclear binding and decay become primary problems. Essentially, all of modern particle physics is later discovered in the attempt to understand nuclear interactions.
1933-1934	Fermi proposes a theory of b decay. His attempts result into first introduction of weak interaction, first explicit use of neutrinos, and first theory of processes in which a "fundamental" particle changes type ($n \to p + e^{-} + \gamma e$). The introduction of new interactions, in addition to the familiar electromagnetic and gravitational interactions, is a very radical step. The idea that the electron (and neutrino) produced in nuclear beta decay are in no way present in the nucleus before the decay is also radically new.
1933-1934	Yukawa speculates about the nature of nuclear forces. Combining relativity and quantum theory, Yukawa tries to describe nuclear interactions by an exchange of new particles between protons and neutrons. From the size of the nucleus, which gives the range of the new interactions, Yukawa concludes that mass of these conjuncture particles (mesons) is about 200 electron masses. This is the beginning of the meson theory of nuclear forces. Nobel Prize awarded to Yukawa in 1949.
1937	A particle of mass about 200 electron masses is discovered in cosmic rays by Neddermeyer and Anderson and by Street and Stevenson. First believed to be Yukawa's meson, it is later recognized to be another entirely new type of particle, now called a muon (see 1946-1947)
1938	Sruckelberg observes that protons do not decay into any combination of electrons, neutrinos or muons or their antiparticles. This stability of the proton cannot be explained in terms of energy and electric charge conservation. He proposed an independent conservation law of massive particles. In contemporary language, this is the conservation law of baryon number or quark number. In 1949 Wigner gives a more explicit formulation, observing that $p \to e^{+} + \gamma$ does not occur.
1941	Measurement of the muon life time (still thought to be Yukawa's meson).
1941	Moller and Pais introduce the term "nucleon" as a generic term for protons and neutrons.
1946-1947	Realization that the cosmic ray particle thought to be the Yukawa meson cannot be any such thing because it does not interact strongly enough as it passes through matter. It is instead a particle with no strong interaction, just like an electron, except that it is more massive and thus more unstable. The

"muon" (μ), the first particle of the second generation to be found, is completely unexpected. It was the first of many unexpected particle discoveries; I.I. Rabi comments (like a customer in a restaurant receiving an unexpected dish), "Who ordered that? " The term "lepton" is introduced as a generic name for objects that do not interact strongly, namely, the electron, muon and neutrinos. Correspondingly, the generic term "hadron" is introduced for particles that do have strong interaction, such as the proton and neutron, and Yukawa's still undiscovered meson.

1947 — Powell and collaborators, using sensitive nuclear emulsions exposed to cosmic radiation, discover another type of particle with a mass little greater than a muon, which does interact strongly. It decays into muon and neutrino. It is Yukawa's strongly interacting meson, now called the pion.

1947 — Development of computational procedures for quantum electrodynamics (QED), the relativistic quantum theory of electromagnetic interactions of electrons, positrons, photons. Introduction of Feynman diagrams.

E.4. The Advent of Accelerator Experiments

1948 — First artificially produced pions observed at the Berkeley synchrocyclotron.

1949 — Fermi and Yang suggest that a π meson is a composite structure of a nucleon and an antinucleon. The idea that a particle could be composite is quite radical in 1949.

1949 — A new type of meson, now called K^+, is discovered by the Bristol group (Powell and collaborators) via its decay, $K^+ \to \pi^+ + \pi^- + \pi^+$. As early as 1944, Leprince-Ringuet and Lheritier had seen smaller than before substantial evidence for the K^+.

1950 — Discovery of the neutral pion π^0 by Panofsky and Steinberger.

1951 — Rochester and Butler discover two new types of particles in tracks produced in a bubble chamber in processes initiated by cosmic rays. They looked for V-like tracks, which can be interpreted as the diverging tracks of two charged particles produced by the decay of a "parent" electrically neutral particle (which leaves no track). The mass of the parent particle is deduced by reconstructing its energy and momentum from those of (known mass) products. The new electrically neutral particles were at first called V particles. The two types found in this way are now called Λ^0 (which was seen by its decays, $\Lambda^0 \to p + \pi^-$) and K^0 (which was seen to decay $K^0 \to \pi^+ + \pi^-$). The Λ^0 was the first particle more massive than a neutron to be discovered.

1952 Discovery of a set of similar mass proton-like, but unstable, "particles" produced in pion-proton scattering, with four different charge states, Δ^{++}, Δ^{+}, Δ^{0}, Δ^{-}. All decay rapidly to a proton or a neutron, plus a pion.

1952 Glaser invents the bubble chamber. The Brookhaven Cosmotron, a 1.3GeV accelerator, starts operation.

1953 Beginnings of a "particle explosion," a true proliferation of "particles." Classification of particles and their decays become a major activity. Particle decays are classified on the basis of their half-life into two groupings. Decays by strong interaction processes have a half-life of order of 10^{-13} seconds and even much longer. These can only be observed when no strong decay is possible. Particles are classified by spin (fermions or bosons) and by the types of interactions in which they participate (hadrons or leptons).

1953 Law of conservation of lepton numbers first stated in paper by Konopinski and Mahmoud.

1953 Introduction by Gell-Mann and Nishijima of a new particle attribute, called "strangeness," to explain the disparity between the copious production of Λ^{s} and K^{s} and their slow decay. It is recognized that particles with strangeness decay by only relatively slow weak interactions, but are produced in pairs of opposite sign of strangeness. The modern interpretation of this property is that Λ contains a strange quark, and the K^{+} contains a strange antiquark (Λ = uds and K^{+} = $s^{-}u$). A strange quark and its antiquark can be produced together in a strong interaction, but each decays only by weak interactions.

1953 V, or neutral, particles are produced by the Cosmotron in Broockhaven, the first accelerator production of V particles.

1953-1957 Scattering of electrons by nuclei reveals that the electric charge inside protons has a distribution of varying density and that even the neutrons have the some internal charge density distribution. Description of this electromagnetic structure of protons and neutrons suggests some kind of internal structure to these objects, though they are still regarded as fundamental particles.

1954 Yang and Mills develop a new class of theories called "non-Abelian gauge theories." This type of theory now forms the basis of the Standard Model.

1956-1957 Lee and Yang observe that the parity P invariance of basic laws (the property that they do not change form when all directions are reflected) and charge-conjugation C invariance (the particle-antiparticle) have never been checked in weak interactions. They suggest that weak interactions may not possess these symmetries, though strong and electromagnetic interaction laws do.

Experiments on b decay by Wu et al. and muon decay indeed show violations of P and C invariance. Nobel Prize was awarded to Lee and Yang in 1957.

1957 Schwinger writes paper proposing unification of weak and electromagnetic interactions.

1957-1959 Separate papers by Schwinger, Bludman and Glashow suggest that all weak interactions are mediated by charged, very massive bosons, later called W^+ and W^-. This idea has a similarity to that of Yukawa when first discussed boson exchange 20 years earlier, when he proposed the pion as the mediator of the strong force. The name W^+ and W^- was first used by Lee and Yang in 1960.

1961 Gell-Mann exploits the patterns of particles of similar mass and spin, but differing charge and strangeness to create a classification scheme [based on the group SU(3)] now called flavor symmetry, for the ever-increasing number of known particles. The scheme predicts a new particle type, Ω^-, whose discovery shortly thereafter gives great validity to this idea. For his earlier work on strangeness, this classification scheme and later work on quarks (see 1964) Gell-Mann was awarded the Nobel Prize in physics in 1969.

1962 Experimental verification that two distinct types of neutrinos (ν_e and ν_μ) exist. This was earlier inferred from theoretical considerations. The experimenters, Lederman, Schwartz, and Steinberger were awarded the Nobel Prize in 1988 for this work.

E.5. Formulation of the Modern View

1964 First tentative introduction of quarks, by Gell-Mann and independently by Zweig. Quarks give a basis in terms of the particle structure for the classification scheme proposed earlier by Gell-Mann. All mesons and baryons are composites of three species of quarks and antiquarks, now called u, d, and s of spin ½ h and with electric charges (2/3, -1/3, -1/3) in units in which the proton charge is 1. The similarly charged d and s quarks are distinguished by the fact that the s carries the "strangeness" quantum number -1, while the d has zero for this quantity. These fractions of a proton or electron charge had never been observed, and so the introduction of quarks was generally treated more as a mathematical explanation of flavor patterns of particle masses (see 1961) than as a postulate of actual physical objects. However, the great simplification from over 100 "fundamental" hadronic particle types to just three makes the idea very attractive. Later, theoretical and experimental developments allow us to now regard the quarks as real physical objects, even though they cannot be isolated.

1964 Stimulated by the repeated pattern of leptons, several papers suggest a fourth quark carrying another flavor to give a similar repeated pattern for the quarks, now seen as flavor generation patterns. Very few physicists take this suggestion seriously at the time. Glashow and Bjorken coin the term "charm" for the fourth (c) quark.

1965 Color charge an additional degree of freedom, is introduced as an essential property of quarks by Greenberg and by Han and Nambu. All observed hadrons are presumably color neutral.

1966 The quark model is accepted rather slowly. Quarks were not observed. In Gasiorowicz's popular summarizing textbook on particle physics, published in 1966, quarks are not mentioned.

1967 Weinberg and Salam separately propose a theory that unifies electromagnetic and weak interactions. It is of the Yang-Mills type (see 1954). The theory requires the existence of a neutral, weakly interacting boson (now called the Z^0) that mediate certain weak interactions that had not been observed at that time. They also predict an additional massive boson called the Higgs Boson, which has not been yet observed, to explain the particle masses. Their idea is mostly ignored; from 1967 to 1971, Weinberg's and Salam's papers, now regarded as the first suggestion of an essential part of the Standard Model, were only quoted five times.

1968-1969 Observations at the SLAC laboratory indicate that in inelastic electron-proton scattering the electrons appear to be bouncing off small dense objects inside the proton. Bjorken and Feynman analyze this data in terms of a model of constituent particles inside the proton, without using the name quark for the constituents. Taylor, Friedman, and Kendall are awarded Nobel Prize in 1990 for this experimental evidence for quarks.

1970 The critical importance of a fourth type of quark in the context of the Weinberg-Salam type theory of weak and electromagnetic interactions (see 1967) is recognized by Glashow, Iliopoulos, and Maiani. A fourth quark allows a theory that has flavor-conserving Z0-mediated weak interactions but not flavor-changing ones. The known absence of the flavor-changing type had led experimenters to ignore a search for the flavor-conserving ones until this theory predicted its presence!

1972 't Hooft and Veltman develop computational tools that allow the Weinberg-Salam theory to be treated beyond simple approximation. Recognition that this theory is well behaved (compared to the Fermi theory, which gives nonsense beyond the simple approximation) leads to growing interest in it.

1972 Definite formulation of a quantum field theory of strong interactions. This theory of quarks and gluons (now part of the Standard Model) is similar in mathematical structure to quantum electrodynamics (QED), hence the name quantum electrodynamics (QED). It is also a Yang-Mills type theory. Quarks are real particles carrying a color charge. Gluons are mass-less quanta of the strong-interaction field, which also carry color charges. This strong interaction theory is first suggested by Fritzsch and Gell-Mann.

1973 Spurred by a prediction of the Weinberg-Salam-Glashow, Iliopoulos, and Maiani theories, the Gargamelle collaboration reanalyzes some old data from CERN and finds indications of weak interactions with no charge exchange (those due to a Z^0 exchange). Here the interplay of theory and experiment is interesting. Before the theoretical prediction of Z^0, everyone just assumed, that no weak-interaction processes existed that conserved both flavor and charge, and the early analysis of the CERN did not really look for them. It was stated that the search would be too difficult because of "background" processes (those from similar-looking, but non-weak-interaction processes). After the theoretical prediction, the experimenters found it was possible to exclude background events. Theoretical bias often decides how hard we think about how to make a certain measurement.

1974 Politzer, a graduate student at Harvard, and Gross, at Princeton with his student Wilczek, discovers that theories such as the color theory of the strong interactions have a special property, now called "asymptotic freedom." The property is necessary to describe correctly the 1968-1969 data on the structure of the proton. No other class of theories has this property.

1974 In a summary talk for a summer conference, Ilipoulos summarizes, for the first time in one single report, the view of physics now called the Standard Model. Only particles containing u, d, and s quarks were known, with the charm quark predicted by the theory, but particles containing it not yet discovered. Most physicists are skeptical about the need for charm, but Glashow declares at an international conference in April that if charm is not found within two years, he would "eat my hat."

E.6. Experimental Verification of the Standard Model

1974 November: An extremely narrow, high peak in event rate, signaling production of a new type of a particle is discovered by Richter and his collaborators in the reaction $e^+ + e^- \rightarrow$ hadrons at SLAC (they call the particle ψ). The same object is discovered by Ting and his collaborators at about the same time at Brookhaven in a proton experiment and called J. Actually they found it earlier but kept it secret while they made various checks and tried to understand what they had seen. The two discoveries are announced on the same day, and so are given equal credit. There is a great deal of excitement about the discovery of this new particle of mass 3.1 GeV, now called J/ψ. There are many wild speculations about what J/ψ is. The most favored idea is that it is a cc' state. If so, states of u'c and d'c (D mesons) must also exist. The 1976 Nobel Prize was shared by Richter and Ting for this discovery.

1976 A neutral, charmed meson, called a D^0 meson, is found by Goldhaber and Pierre. The mass is 1.87 GeV, its quark content is u'c, and the observed decay is $D^0 \rightarrow K^- + \pi^+$, as is predicted for a charmed meson. It is produced in combination with another such particle D^0'nmade from c'u, which decays to $K^+ + \pi^-$. These are the first mesons that possess charm or anticharm but not both. The experimental results agree dramatically with the theoretical predictions and confirm the interpretation of the J/ψ as a cc' state.

1976 A new charged lepton called τ (tau), of mass about 1.78 GeV, is recognized by Perl and collaborators at SLAC. It is totally unexpected. The production of particle-antiparticle pairs of this lepton at almost the same energy as the charm quark and antiquark production threshold had led to some initial confusion in the interpretation of the J/ψ discovery. Only after the discoveries of the τ lepton and the D mesons were sorted out was the data completely understood. This is a strange repetition of history: Just as the discovery of a first generation meson (pion) was confused by the unexpected appearance of second generation lepton (muon), so now the interpretation of the second generation mesons (J/ψ and D) is confused by the appearance of equally unexpected third generation lepton (τ).

The fact that the e and μ leptons possess different associated strongly suggests that the τ lepton also possesses an associated neutrino, giving six leptons in all.

1977 Discovery of particles, called Y (Upsilon), containing yet another quark (and its antiquark), by Lederman and collaborators at Fermilab. It is called the "bottom" quark, charge -1/3. It gives added impetus to the search for a sixth quark ("top"), so that the number of quarks would equal the number of leptons and the third repeat of the pattern of particles in the "Standard Model" (third generation) be complete.

1978 Observation of the effect of Z^0-mediated weak interactions in the scattering of polarized electrons from deuterium, in an experiment at SLAC led by Prescott and Taylor. The experiment qualitatively and quantitatively confirms a key prediction of the Standard Model.

1979 PETRA, a colliding-beam facility at DESY in Hamburg, Germany, studies events with clusters of hadrons seeking evidence for gluon-induced effects predicted in QCD. The reaction $e^+ + e^- \rightarrow q + q'$ with subsequent production of additional quark-antiquark pairs yields two clusters of hadrons moving in the direction of initial quark and initial antiquark. A smaller number of events has three clusters, strong evidence for the existence of a gluon radiated by initial quark or antiquark ($e^+ + e^- \rightarrow q + q' +$ gluon). The result fit the patterns predicted.

1979 The Nobel Prize is awarded to Glashow, Salam, and Weinberg for their role in the development of the electroweak theory, four years before the observation of the W^+, W^- and Z^0 bosons predicted by their theory, but after the discovery of weak neutral currents, that is, processes mediated by Z^0 boson. This indirect evidence was sufficiently convincing to physicists that the theory must be right.

1983 The W^+, W^- and Z^0 intermediate bosons demanded by the electroweak theory are observed by two experiments using the CERN synchrotron, converted into a proton-antiproton collider by van der Meer and his team. The observations are in excellent agreement with the theory. The spin of the W^+ and W^- could be measured: It is 1h', as required by the Standard Model. The 1984 Nobel Prize was awarded to Rubbia and van der Meer for this work.

1989 Studies at CERN and SLAC of properties for the Z^0 in e^+ e^- collisions establish the existence of exactly three low-mass neutrino species with standard interactions, strongly implying that only three families of fundamental particles exist.

1989-1993 Millions of Z^0's produced by LEP collider at CERN permit study of the properties of the Z^0 and its decay products in great detail and provide precise tests of the Standard Model.

1989-1995 Measurement of the mass and width of the W^+ and W^-, at CERN and at Fermilab, together with the properties of the Z^0, provide further tests of the electroweak aspects of the Standard Model. Other tests, for example, the energy dependence of the rate of producing multiple particle clusters, provide equally strong support for the strong interaction (QCD) aspect of the Standard Model.

1995 After 18 years of searching at many accelerators, the CDF and DO experiments at Fermilab discover the top quark at the mass of 175 GeV. No one understands why the mass is so different from the other five quarks. In fact, no one understands any of the patterns of quark and lepton masses. The Standard Model can fit them but does not predict them.

1998 Evidence accumulates from a number of experiments on neutrinos from the sun and those produced in the atmosphere from cosmic rays, that it may be necessary to extend the Standard Model by adding right-handed neutrinos and very small neutrino masses.

E.7. Advances in relativity theory at the onset of 21st Century

2002 Kadakia, an electrical engineer by profession suggested that light energy and various other forms of energy in electromagnetic spectra behaves strictly as waves whose energy is proportional to frequency. He explained nature of light as being a wave and explained results of all the experiments that were performed to prove particle nature of light and all the phenomena that exhibited light as photon particles by wave theory of light. He gave name **Skylativity®** theory to the principles of wave theory of light.

2009 Kadakia establishes that the speed of light C is variable and depends on frame of reference. The observed red-shift or Z-shift of the radiation energy from celestial objects is an apparent or virtual shift because of Doppler's effect.

2009 Kadakia proves that mass of stars such as the sun remains constant inspite of billions of Mega watts of radiation energy emission per second from its surface.

F

APPENDIX: References and Suggestions for Further Reading

F.1. References

[1] H.A., Einstein, A., Minkowski, H., and Weyl, H., The Principle of Relativity (Methuen, London, 1923).

[2] Robert Mills, " SPACE TIME and QUANTA - An Introduction to Contemporary Physics," W. H. Freeman and Company New York, 1994.

[3] Anthony French & Edwin Taylor, "An Introduction to Quantum Physics," W.W.Norton & Company, 1978.

[4] John Peacock, "Cosmological Physics," Cambridge University Press, 1999, pp. 3-145

[5] Michael Barnett, Henry Muhry and Helen Quinn, "The Charm of Strange Quarks: Mysteries and Revolutions of Particle Physics," AIP Press, 2000, pp. 211-291.

[6] Arthur Beiser, "Concepts of Modern Physics," McGraw-Hill, Inc., 1981.

[7] Max Born, "Einstein's Theory of Relativity," Dover Publications, Inc. 1962.

[8] James Rutherford, Gerald Holton, Fletcher Watson, "The Project Physics Course", Holt, Rinehart and Winston, Inc. 1972.

[9] Peter Bergmann, "Introduction to the Theory of Relativity with a Forward by Albert Einstein," Dover Publications, Inc. 1976.

[10] Dennis Derickson, "Fiber Optic Test and Measurement," Prentice Hall PTR, Inc. 1998.

[11] N. Wanadoo, "General Relativity: A Brief Explanation of the Fundamental Ideas," Web

[12] R Highfield, "Searching For the Dark Force," CALGARY HERALD, August 23, 2003

[13] Anthony Browne,"Theory of Relativity Could be all Wrong," CALGARY HERALD, August 09, 2002

[14] Brian Greene," The Elegant Universe: Super Strings, Hidden Dimensions and the Quest for the Ultimate Theory," W. W. Norton & Company, Inc. 1999.

[15] Roger Penrose, " SIX NOT SO EASY PIECES Einstein's Relativity Symmetry, and Space-Time RICHARD P. FEYNMAN," Addison Wesley Publishing Company, Inc., 1997.

[16] Tim Folger, "At the speed of Light – What if Einstein was wrong, "DISCOVER April, 2003 Pages 34-41.

[17] Roger Freedman & William Kaufmann III, "UNIVERSE," W. H. Freeman and Company-New York, 2001

[18] University of Northern Colorado Physics Class Project, Report by Steve Clark, Spring Semester 2005.

[19] Hans Ohanian & Remo Ruffini, "Gravitation and Space-time," W.W.Norton & Company, 1994.

[20] Charles Misner, Kip Thorne and John Wheeler, "Gravitation," W. H. Freeman and Company-New York, 1973.

[21] Kip Thorne, "Black Holes & Time Warps Einstein's Outrageous Legacy," W.W.Norton & Company, 1994.

[22] Raymond Serway & John Jewett, Jr., "Physics for Scientists and Engineers with Modern Physics," Saunders College Publishing, 2008.

[23] //www.yahooligans.com/Science_and_Nature/Physical_Sciences/Physics/Relativity

[24] //www.humboldt1.com/~gralsto/einstein/1905.html and //whyfiles.org/052einstein/index.html

[25] //www.geoastro.de/astro/temperature.htm

[26] Chandrasekhar, S. "An Introduction to Study of Stellar Structure", the University of Chicago, Chicago, Illinois

[27] John Beacom & Nicole Bell, "Do solar neutrinos decay?" NASA/Fermilab Astrophysics Center, April 09, 2002.

[28] //www.geoastro.dc/astro/temperature.htm

[29] John Bahcall, "How the Sun Shines", Beam Line, Winter 2001.

[30] Celal Ceyhan, "The Interior of the Sun", Umea University, January, 2006.

[31] en.wikipedia.org/wiki/Mean_free_path "Mean free path in kinetic theory

[32] http://physicsworld.com, "Paul Dirac: the pursuit soul in physics" Feb 1, 1998.

[33] http://en.wikipedia.org/wiki/Dirac_equation

[34] http://newsroom.ucr.edu/news_item.html, "French-German Group Verifies High-Temperature Superconductivity Theory Proposed by UCR Physicist".

[35] Guo-Meng Zhao, "The resistive transition and Meissner effect in carbon nano-tubes: Evidence for Quasi-one-dimensional superconductivity above room temperature", Department of Physics and Astronomy, California State University at Los Angeles, Los Angeles, CA 90032, USA

[36] http://www.jpl.nasa.gov/news/features.cfm, " All days are not created equal".

[37] Shailesh Kadakia, "Light is a wave: Revisiting outcome of light's particle nature experiments", March 2, 2010.

[38] Shailesh Kadakia, "Are light waves different from Electromagnetic waves: New model?" March 2, 2010.

[39] Lyndon Taylor, "Introductory Materials for Electrical Engineers" 1978.

[40] Oran Brigham,"The Fast Fourier Transform," Prentice Hall, Inc., 1974

F.2. Articles

Particle Physics Articles

1. E. Bloom and G. Fieldman, "Quarkonium," Scientific American (May 1982) p. 66.
2. Boslough, "World within the Atom," National Geographic (May 1985) p. 634
3. D. Cline, C. Rubbia, and S. van der Meer, "The Search for Intermediate Vector Bosons," Scientific American (March 1982) p. 48
4. D. Cline, "Low Energy Ways to Observe High Frequency Phenomenon," Scientific American (September 1984)
5. F. E. Close and P. R. Page, "Glueballs," Scientific American (November 1998) p. 52
6. J. W. Cronin, S.P. Swordy and T.K. Gaissser. "Cosmic Rays at the Energy Frontier," Scientific American (January 1997) p. 44
7. M.J. Duff, "The Theory Formerly Known as Strings," Scientific American (September 1998) p. 64
8. H. Georgi, "A Unified Theory of Elementary Particles and Forces," Scientific American (April 1981) p. 48
9. C. Grab, H. Breuker, H. Drevermann, and A.A. Rademaker, "Tracking and Imaging of Elementary Particles," Scientific American (August 1991) p. 42

10. H. Harari, "The Structure of Quarks and Leptons," Scientific American (April 1983) p. 56
11. R. C. Howis and H. Kragh, "P.A.M. Dirac and the Beauty of Physics," Scientific American (May1993) p. 62
12. M. Kaku, "Into the Eleventh Dimension," New Scientist (January 18, 1997) p. 32
13. A.M. Litke and A.S. Schwarz, "The Silicon Microchip Detector," Scientific American (February 1994) p. 56
14. C. Mann, "Armies of Physicist Struggle to Discover Proof of a Scot's Brainchild," Smithsonian (March 1989) pp. 106-117.
15. C. Quigg, "Elementary Particles and Forces," Scientific American (April 1985) p. 84
16. C. Quigg, "Top-ology," Physics Today (May 1997) p. 20.
17. H.R. Quinn and M.S. Witherell, " The Asymmetry Between Matter and Antimatter," Scientific American (October 1998) p. 76
18. C. Sutton, "Subatomic Forces," New Scientist (February 11, 1989) p. 1.
19. C. Sutton, "Four Fundamental Forces," New Scientist (November 19, 1988) p. 1.
20. C. Sutton, "The Secret Life of the Neutrino," New Scientist (January 14, 1988) p. 53.
21. M. Veltman, "The Higgs Boson," Scientific American (November 1986) p. 76.
22. S. Weinberg, "The Discovery of Subatomic Particles," Scientific American (1983) p. 206
23. S. Weinberg, "A Unified Physics by 2050?," Scientific American (December 1999)
24. D.H. Weingarten, "Computing Quarks," Scientific American (February 1996) p. 116
25. E. Witten, "Duality, Spacetime and Quantum Mechanics," Physics Today (May 1997) p. 28
26. E. Witten, "Reflections on the Fate of Spacetime" Physics Today (April 1996) p.

Accelerator Physics Articles

1. Winick, "Synchroton Radiation," Scientific American (November 1987) p. 88

Cosmology Articles

1. J. N. Bahcall and F. Halzen, "Neutrino Astronomy; The Sun and Beyond," Physics World (September 1996) p. 41.
2. N. Bahcall, J.P. Ostriker, S. Perlmutter, and P.J. Steinhardt, "The Cosmic Triangle Revealing The State of The Art Universe," Science (May 1999) p. 1481.
3. S.G. Brush , " How Cosmology Became a Science," Scientific American (August 1992) p. 34.
4. M.A. Bucher and D.N. Spergel, "Inflation in a Low Density Universe," Scientific American (January 1999) p. 62.
5. Roger A. Freedmann, "The Expansion Rate and the Size of the Universe," Scientific American (July 1993) p. 70.
6. R. Gore, "The Once and Future Universe," National Geographic (June 1983) p. 704.
7. C.J. Hogan, R.P. Kishner, and N.B. Suntzeff, "Surveying Space time with Supernovae," Scientific American (January 1999) p. 46.
8. R .Irion, " The Lopsided Universe," New Scientist (February 1999) p. 26.

9. L.M. Krauss and G.D. Starkman, "The Fate of the Life in the Universe," Scientific American (November 1999) p. 58.
10. L.M. Krauss, "Cosmological Antigravity," Scientific American (January 1999) p. 52.
11. D.E. Osterbroack, J.A. Gwinn, and R.S. Brashear, "Hubble and the Expansion of the Universe," Scientific American (November 1992) p. 30.
12. Schramm and G. Steigmann, "Particle Accelerators Test Cosmological Theory," Scientific American (June 1988) p. 66.
13. C. Sutton, "Cosmic Changelings," New Scientist (March 16, 1996) p. 28.
14. M.S. Turner, "Cosmology: Going Beyond the Big \Bang," Physics World (September 1996) p. 31.

F.3. Books Particle Physics Books

1. R.K. Adair, The Great Design: Particles, Fields and Creation (University Press 1987)
2. I. Asimov, Atom: Journey across the Subatomic Cosmos (Truman Talley Books, 1991)
3. L. Brown and L. Hoddeson, Eds., The Birth of Particle Physics (Cambridge University Press, 1983).
4. B. Bunch, Reality's Mirror: Exploring the Mathematics of Symmetry (Wiley Science Editions, 1980)
5. R.N. Kahn and G. Goldhaber, The Experimental Foundation of Particle Physics (Cambridge University Press, 1989).
6. N. Calder, The Key to the Universe (Penguin Books, 1978)
7. R. Carrigan and P. Trower, Ed., Particles and Forces at the Heart of Matter, Readings from Scientific American (Freeman 1990).
8. C. Caso, et. Al., (Particle Data Group) Particle Physics Booklet (Springer Verlag, 1998)
9. F. Close, M. Marten, and C. Sutton, The Particle Explosion (Oxford University Press, 1987).
10. R.P. Crease, C.C. Mann, and T. Ferris The Second Creation: Makers of the Revolution in 20th Century Physics (Rutgers Univ. Press, 1996).
11. P.C. Davies, The Forces of Nature (Cambridge University Press, 1986).
12. B. Devine and F. Wilczek, Longing for the Harmonics (Norton, 1988)
13. G. Fraser, The Particle Century (Institute of Physics, 1998).
14. G. Fraser, The Quark Machines: How Europe Fought the Particle Physics War (Institute of Physics, 1997).
15. G. Fraser, E. Lillestoel, and I. Sellevag, The Particle Search for Infinity: Solving the Mysteries of Universe (Facts on File 1995).
16. H. Fritzsch, The Creation of Matter: The Universe from Beginning to End (Basic Books, 1984. Translated from German 1981).
17. S. Glashow, Interactions: A Journey Through the Mind of a Particle Physicist and the Matter of this World (Warner, 1988)
18. J. R. Gribbin, Q is for Quantum: An Encyclopedia for Particle Physics (Free Press, 1999)

19. J.R. Gribbin, The Search for Super Strings, Symmetry and the Theory of Everything (Little Brown, 1999)
20. A.H. Guth and A.P. Lightman, The Inflationary Universe: The Quest for a New Theory of Cosmic Origin (Perseus Press, 1998)
21. M. Kaku, Hyperspace: A Scientific Odyssey through Parallel Universes, Time Warps and the 10th Dimension (Oxford University Press, 1994)
22. G. Kane, The Particle Garden (Addision-Wesley Publishing Company, 1995).
23. L.M. Krauss, Fear of Physics (Basic Books, 1993)
24. L.M. Lederman, The God Particle: If the Universe is the Answer, What is the Question? (Hougton Mifflin, 1993)
25. J.H. Mauldin, Particles in Nature: The Cosmological Discovery of the New Physics (Tab Books, 1986).
26. Y. Ne'eman and Y. Kirsh, The Particle Hunters (Cambridge University Press, 1986)
27. L.B. Okun, α,β,γ…..Z, A Primer in Particle Physics (Harwood Academic Publishers, 1987)
28. Pagels, Perfect Symmetry (Simon & Schuster, 1985).
29. A. Pais, Inward Bound---Of Matter and Forces in the Physical World (Oxford University Press, 1988).
30. M. Riordan, The Hunting of the Quark (Simon & Schuster, 1987).
31. M. Riordan and D.N> Schramm, Shadows of Creation: Dark Matter and the Structure of the Universe (Scientific American Library, Freeman, 1991).
32. C. Schwarz, A Tour of the Subatomic Zoo (American Institute of Physics, 1992)
33. E. Segre, From X-rays to Quarks: Modern Physicists and Their Discoveries (W. H. Freeman, 1980).
34. G. Smoot and K. Davidson, Wrinkles in Time (Avon Books, 1994).
35. N. Solomey, The Elusive Neutrino: A Subatomic Detective Story (Scientific American Library Series, 1997).
36. C. Sutton, Spaceship Neutrino (Cambridge University Press, 1992)
37. S.J. Traweek, Beamtimes and Lifetimes (Harvard University Press, 1986).
38. J.S. Trefil, From Atoms to Quarks (Doubleday, 1994)
39. P. Watkins, Story of the W and Z (Cambridge University Press, 1986).
40. S. Weinberg, Dreams of a Final Theory (Pantheon, 1992).
41. A. Zee, Fearful Symmetry (Macmillan, 1986).

Cosmology Books

1. J.D. Barrow and J. Silk, The Left Hand of Creation (Oxford Univ. Press, 1994).
2. F. Close, The Cosmic Onion (Morrow, 1988)
3. T. Ferris, Coming of Age in the Milky Way (Morrow, 1988).
4. D. Goldsmith, Einstein's Greatest Blunder? The Cosmological Constant and Other Fudge Factors in the Physics of Universe (Harvard University Press, 1995).
5. B. Greene, The Elegant Universe: Superstring, Hidden Dimensions, and the Quest for the Ultimate Theory (W.W. Norton, 1999).
6. S.W. Hawking, A Brief History of Time (Bantam, 1988).

7. L.M. Krauss, The Fifth Essence: The Search for Dark Matter in the Universe (Basic Books, 1989).
8. L.A. Marchall, The Supernova Story (Plenum Press, 1988).
9. H. Pagels, The Cosmic Code (Simon & Schuster, 1982).
10. M. Roos, Introduction to Cosmology (Wiley, 1994).
11. J. Silk, The Big Bang (Freeman , 1989).
12. S. Weinberg, The First Three Minutes (Basic Books, 1977).
13. Jayant V Narlikar, R.J. Tayler, W. Davidson and M.A. Ruderman , ASTROPHYSICS,
W.A. Benjamin, London, 1969
14. Jayant V Narlikar and Fred Hoyle, ACTION AT A DISTANCE IN PHYSICS AND COSMOLOGY, W.H. Freeman and Company, San Francisco, 1974
15. Jayant V Narlikar GANIT ANI VIDNYAN (Mathematics and Science in Marathi) Mumbai Marathi Grantha Sangrahalaya, Bombay, 1975
16. Jayant V Narlikar, GENERAL RELATIVITY AND COSMOLOGY, The Mc Millan Company of India Ltd., New Delhi, 1978.
17. Jayant V Narlikar, THE LIGHTER SIDE OF GRAVITY (in Russian: translation of 12A by G.I. Blinnikova) Mir Publications, Moscow, 1985
18. Jayant V Narlikar and T. Padmanabhan GRAVITY, GAUGE THEORIES AND QUANTUM COSMOLOGY, D. Reidel, Dordrecht, 1986
19. Jayant V Narlikar, THE LIGHTER SIDE OF GRAVITY, (in Japanese: translation of 12A by K. Nakamura) Nikkei Science, Inc, Tokyo, 1986)
20. Jayant V Narlikar, UNE GRAVITATION SANS GRAVITE, (in French: translation of 12 A by Jean-Pierre Maury) Payot, Paris, 1986.
21. Jayant V Narlikar, LE ESTRUCTURA DEL UNIVERSO, (in Spanish : Translation of 4A by Artur Klein) Alienza Editorial, Madrid, 1987
22. Jayant V Narlikar, FENS'MENOS VIOLENTOS EN EL UNIVERSO, (in Spanish: translation of 11A by Artur Klein) Alienza Editorial, Madrid, 1987.
23. Jayant V Narlikar, THE PRIMEVAL UNIVERSE, Oxford University Press, Oxford, 1988.
24. Jayant V Narlikar, HIGHLIGHTS IN GRAVITATION AND COSMOLOGY (Proceedings of ICGC-87; co-editors: B.R. Iyer, A. Kembhavi, and C.V. Vishveshwara) Cambridge University Press, Cambridge, 1988
25. Jayant V Narlikar, THE FRONTIER BETWEEN PHYSICS AND ASTRONOMY, (IIT Madras Series in Science and Engineering) Macmillan India Limited, Delhi, 1989
26. Jayant V Narlikar, VISHVANI UTPATTI (Origin of the Universe in Gujarati : translated by J.J. Rawal) Parichaya Trust, Bombay, 1989
27. Jayant V Narlikar, THE PHYSICS ASTRONOMY FRONTIER, (in Chinese: translation of 8A by Xiang Tao He) Beijing, 1989
28. Jayant V Narlikar, THE LIGHTER SIDE OF GRAVITY, (in Greek: translation of 12 A by Andreas I. Kathetas) Bibliotheka Groxalia, Athens, 1989
29. Jayant V Narlikar, FROM BLACK CLOUDS TO BLACK HOLE, (in Russian) Energo-Atomizdat, Moscow, 1989
30. Jayant V Narlikar, BRAHMAND KI KUCHH JHALAKEN (A Few Glimpses of the Universe in Hindi) Madhya Pradesh Hindi Granth Akademi, Bhopal, 1990

31. Jayant V Narlikar, VIDNYANACHI GARUDA JHEP (The Leap-frogging of Science in Marathi) Shrividya Prakashan, Pune, 1990
32. Jayant V Narlikar, VIDNYAN ANI VAIDNYANIK (The Science and the Scientists in Marathi)
Shrividya Prakashan, Pune, 1990.
33. Jayant V Narlikar, TARONKI JEEVAN GATHA (The Life Story of Stars in Hindi) Read and Learn Series of NCERT, Delhi, 1990
34. Jayant V Narlikar, FROM BLACK CLOUDS TO BLACK HOLES, (in Japanese : translation of 18A by Tadaoki Yoneyama and Kenji Urata) Maruzen Co. Ltd., Tokyo, 1990
35. Jayant V Narlikar, VISHVAYAN (The Cosmic Spaceship in Kannada : translation of 21A by G.T. Narayana Rao), 1991
36. Jayant V Narlikar, VIDNYANACHE RACHAYITE (The Creators of Science in Marathi : series of 12 booklets on famous scientists by R.V. Sovni) Oxford University Press, Bombay, 1991
37. Jayant V Narlikar, GANITATALYA GAMATI-JAMATI (Recreations in Mathematics in Marathi) Manovikas Prakashan, Bombay 1992
38. Jayant V Narlikar, VISHVACHI RACHANA, (The Origin of the Universe in Marathi) Wiley Eastern, Delhi 1992
39. Jayant V Narlikar, ASTROFISICA, (Astrophysics) in Italian) Jaca Book, 1992
40. Jayant V Narlikar, PHILOSOPHY OF SCIENCE : Perspectives from Natural and Social Sciences (Co-editors : Indu Banga and Chhanda Gupta) Indian Institute of Advanced Study, Simla and Munshiram Manoharlal Publishers, Delhi, 1992
41. Jayant V Narlikar, INTRODUCTION TO COSMOLOGY (2nd Edition) Cambridge University Press 1993. Also published by Foundation Books, New Delhi, 1993
42. Jayant V Narlikar, VISHVACHI SAHJAL (A Journey Through the Universe in Marathi) National Book Trust's Nehru Bal-Pustakalaya, New Delhi, 1988.
43. Jayant V Narlikar, GRAVITATION AND RELATIVITY : AT THE TURN OF MILLENNIUM (Co-editor : Naresh Dadhich) Proceedings of the GR-15 Conference held at IUCAA, Pune, India during December 16-21, 1997
44. Jayant V Narlikar, LZEJSZA STRONA GRAWITACJI (in Polish: translation of 73A by Jerzy Lewinski) Wydawnictwo Amber Sp.z.o.o.1998 \
45. Jayant V Narlikar, ATITHI (in Malayalam : translation of 14A by K. Krishnan Kutty) D.C. Books, Kottayam, 1998
46. Jayant V Narlikar, MOTION AND GRAVITY, Book in the Exploratory Series, Shekhar Phatak & Associates, Pune, 1999
47. Jayant V Narlikar and Ajit K. Kembhavi, QUASARS AND ACTIVE GALACTIC NUCLEI : AN INTRODUCTION, Cambridge University Press, Cambridge, 1999
48. Jayant V Narlikar, SEVEN WONDERS OF THE COSMOS Cambridge University Press, Cambridge, 1999
49. Jayant V Narlikar, Fred Hoyle and Geoffrey Burbidge, A DIFFERENT APPROACH TO COSMOLOGY Cambridge University Press, Cambridge 2000
50. Jayant V Narlikar, MAHA VISWE MAHAKASHE (in Bengali : translation of a collection of science essays on astronomy and astrophysics) Shaibya Prakashan Bibhag, Calcutta, 2000

51. Jayant V Narlikar, LAS SIETE MARAVILLAS DEL COSMOS (in Spanish : translation of 88A by Dulcinea Otero Pineiro and David Galadi-Enriquez) Cambridge University Press, Cambridge, 2000

52. Jayant V Narlikar and Mangala J. Narlikar, FUN AND FUNDAMENTALS OF MATHEMATICS, Universities Press (India) Ltd., Hyderabad, 2001

53. Jayant V Narlikar, DIE SIEBEN WUNDER DES UNIVERSUMS (in German: translation of 88A by von Helmut Mennicken, Roger & Bernhard GmbH & Co., Germany, 2001

54. Jayant V Narlikar, ANTARAL ANI VIDNYAN (Space and Science in Marathi) Navachaitanya Prakashan, Mumbai, 2002

55. Jayant V Narlikar, AN INTRODUCTION TO COSMOLOGY, (Third Edition) Cambridge University Press, Cambridge, 2002

56. Jayant V Narlikar, THE SCIENTIFIC EDGE, Penguin Books India (P) Ltd., New Delhi, 2003

57. Jayant V Narlikar, LE SETTE MERAVIGLIE DEL COSMO, (in Italian : translation of 88A)
Codice Edizioni, Torino, 2004

58. Jayant V Narlikar, AVAKASHI SANDESH-VAHAK (in Gujarati : translation of 14A by Bharat Pathak), Gurjar Grantha Ratna Karyalaya, Ahmedabad, 2004

59. Jayant V Narlikar, BLACK HOLES, National Book Trust, India, New Delhi, 2005

60. Jayant V Narlikar, TALES OF THE FUTURE, Witness Books, Delhi, 2005

61. Jayant V Narlikar, HOW & WHY IN ASTRONOMY, Marathi Vidnyan Parishad, Mumbai, 2005

62. Jayant V Narlikar, FUNDAMENTALS OF SOLAR ASTRONOMY, Book in the Astronomy and Astrophysics Series by A. Bhatnagar and W. Livingston, World Scientific Publishing Co. Pte. Ltd., Singapore, 2005

63. Jayant V Narlikar, BHARAT KI VIGYAN-YATRA, (in Hindi : translation of 106A by Vinod Kumar Mishra) Prabhat Prakashan, New Delhi, 2005

64. Jayant V Narlikar, AN INVITATION TO ASTROPHYSICS Book in the Astronomy and Astrophysics Series by T. Padmanabhan, World Scientific Publishing Co. Pte. Ltd., Singapore, 2006

65. Jayant V Narlikar, FIND A HOTTER PLACE! A HISTORY OF NUCLEAR ASTROPHYSICS, Book in the Astronomy and Astrophysics Series by L.M. Celnikier, World Scientific Publishing Co. Pte. Ltd., Singapore, 2006

66. Jayant V Narlikar, CURRENT ISSUES IN COSMOLOGY, (Co-editor : Jean-Claude Pecker) Cambridge University Press, Cambridge, 2006

67. Jayant V Narlikar, NABHAT HASARE TARE, (The smiling stars in the sky), (Co-authors : Ajit Kembhavi and Mangala Narlikar in Marathi) Rajhans Prakashan, Pune, 2008

68. Jayant V Narlikar and Geoffrey Burbidge, ACTS AND SPECULATIONS IN COSMOLOGY, Cambridge University Press, Cambridge, 2008

69. Jayant V Narlikar, VIDNYAN ANI MANAVACHA JEEVAN SANGHARSH (Science and Human Survival in Marathi) Deccan Education Society, Pune, 1979

70. Jayant V Narlikar, STRUKTURA WSZECHS'WIATA (in Polish: translation of 4A by Adam Mazurkiewicz) Panstwowe Wydawnictwo Naukowe, Warshaw 1985

General Index

A

B

C

D

E

F

M

N

O

P

Q

R

S

T

U

V

W

X

Z